Artsiom Lapanik

Liquid crystal systems for microwave applications

Artsiom Lapanik

Liquid crystal systems for microwave applications

Single compounds and mixtures for microwave applications Dielectric, microwave studies on selected systems

Südwestdeutscher Verlag für Hochschulschriften

Impressum/Imprint (nur für Deutschland/ only for Germany)
Bibliografische Information der Deutschen Nationalbibliothek: Die Deutsche Nationalbibliothek
verzeichnet diese Publikation in der Deutschen Nationalbibliografie; detaillierte bibliografische
Daten sind im Internet über http://dnb.d-nb.de abrufbar.
 Alle in diesem Buch genannten Marken und Produktnamen unterliegen warenzeichen-, marken-
oder patentrechtlichem Schutz bzw. sind Warenzeichen oder eingetragene Warenzeichen der
jeweiligen Inhaber. Die Wiedergabe von Marken, Produktnamen, Gebrauchsnamen,
Handelsnamen, Warenbezeichnungen u.s.w. in diesem Werk berechtigt auch ohne besondere
Kennzeichnung nicht zu der Annahme, dass solche Namen im Sinne der Warenzeichen- und
Markenschutzgesetzgebung als frei zu betrachten wären und daher von jedermann benutzt
werden dürften.

Verlag: Südwestdeutscher Verlag für Hochschulschriften Aktiengesellschaft & Co. KG
Dudweiler Landstr. 99, 66123 Saarbrücken, Deutschland
Telefon +49 681 37 20 271-1, Telefax +49 681 37 20 271-0
Email: info@svh-verlag.de
Zugl.: Darmstadt, TU, Diss,, 2009

Herstellung in Deutschland:
Schaltungsdienst Lange o.H.G., Berlin
Books on Demand GmbH, Norderstedt
Reha GmbH, Saarbrücken
Amazon Distribution GmbH, Leipzig
ISBN: 978-3-8381-1292-3

Imprint (only for USA, GB)
Bibliographic information published by the Deutsche Nationalbibliothek: The Deutsche
Nationalbibliothek lists this publication in the Deutsche Nationalbibliografie; detailed
bibliographic data are available in the Internet at http://dnb.d-nb.de.
 Any brand names and product names mentioned in this book are subject to trademark, brand
or patent protection and are trademarks or registered trademarks of their respective holders.
The use of brand names, product names, common names, trade names, product descriptions
etc. even without a particular marking in this works is in no way to be construed to mean that
such names may be regarded as unrestricted in respect of trademark and brand protection
legislation and could thus be used by anyone.

Publisher: Südwestdeutscher Verlag für Hochschulschriften Aktiengesellschaft & Co. KG
Dudweiler Landstr. 99, 66123 Saarbrücken, Germany
Phone +49 681 37 20 271-1, Fax +49 681 37 20 271-0
Email: info@svh-verlag.de

Printed in the U.S.A.
Printed in the U.K. by (see last page)
ISBN: 978-3-8381-1292-3

Copyright © 2010 by the author and Südwestdeutscher Verlag für Hochschulschriften
Aktiengesellschaft & Co. KG and licensors
All rights reserved. Saarbrücken 2010

Preface

In this work dielectric and microwave investigations on different nematic liquid crystalline systems are presented. New knowledge about the influence of the chemical structure of molecules on the microwave performance was obtained. This information was used for the preparation of new, optimized nematic mixtures with high values of tunability and low dielectric losses in the microwave region. These new mixtures can be used for producing new devices like phase shifters or varactors. The dielectric measurements of investigated mixtures show that there is no direct influence of dielectric properties in the low frequency region on the GHz region. The influence of different structural fragments of molecules on their properties in the dielectric and microwave region was studied.

Table of Content

Introduction _____ 5
1. Liquid Crystals _____ 8
    1.1 Director and Order Parameter _____ 8
    1.2 Nematics _____ 9
    1.3 Smectics _____ 10
    1.4 Electrooptical Effects _____ 11
    1.5 Surface Stabilized Ferroelectric Liquid Crystals _____ 12

2. Structure of molecules and physical properties of Liquid Crystals _____ 14
    2.1 Thermostability of mesophases _____ 14
        2.1.1 Influence of R_1 and R_2 _____ 15
        2.1.2 Influence of Core _____ 17
        2.1.3 Influence of Bridge fragments _____ 19
        2.1.4 Influence of Lateral substituents _____ 20
    2.2 Thermal stability of Liquid Crystal mixtures _____ 22
    2.3 Influence of structural elements on the optical birefringence __ 22
    2.4 Influence of structural elements on the dielectric properties __ 24
        2.4.1 Liquid Crystals with low dielectric anisotropy _____ 24
        2.4.2 Liquid Crystals with high dielectric anisotropy _____ 25
    2.5 Magnetic anisotropy _____ 26
    2.6 Dielectric properties of Liquid Crystals _____ 27
        2.6.1 Dielectric spectroscopy _____ 28
        2.6.2 Dielectric modes in nematic Liquid Crystals _____ 31

3. Experimental part _____ 33
    3.1 Electro-optical setup _____ 33
        3.1.1 Switching time measurements _____ 34
        3.1.2 Spontaneous polarization measurements _____ 35
        3.1.3 Tilt angle measurements _____ 35
        3.1.4 Polarizing microscopy _____ 36
        3.1.5 Preparation of Liquid Crystal cells _____ 37
        3.1.6 Thickness of the cells _____ 38
    3.2 Geometry of microwave measurements _____ 39
        3.2.1 Nematics _____ 39
        3.2.2 Ferroelectric Liquid Crystals _____ 39
    3.3 Dielectric measurements _____ 40
    3.4 X-Ray measurements _____ 41
    3.5 Investigated mixtures, preparation _____ 42
        3.5.1 Additives _____ 44
        3.5.2 Base matrixes _____ 49
            3.5.2.1 NCS Matrix _____ 49
            3.5.2.2 5CB Matrix _____ 50
            3.5.2.3 Tolane Matrix _____ 50
        3.5.3 Overview of the prepared mixtures based on the NCS, 5CB
            and Tolane matrixes _____ 51
        3.5.4 Mixtures with high optical anisotropy _____ 55
        3.5.5 Mixture with negative dielectric anisotropy _____ 59

4 Results and Discussion _____ 60
   4.1 Ferroelectric LC's – X-Ray and microwave measurements _____ 60
      4.1.1 X-Ray diffraction of FLC's _____ 60
      4.1.2 Microwave measurements of a FLC _____ 64
   4.2 Nematic mixtures at MHz region _____ 67
      4.2.1 Mixtures based on the NCS matrix _____ 67
      4.2.2 Mixtures based on the cyano-biphenyl matrix _____ 73
      4.2.3 Mixtures based on the Tolane matrix _____ 79
   4.3 Nematic mixtures at 30GHz _____ 83
      4.3.1 Mixtures based on the NCS matrix _____ 83
      4.3.2 Mixtures based on the cyano-biphenyl matrix _____ 87
      4.3.4 Mixtures based on the Tolane matrix _____ 90
   4.4 Nematic mixtures at 38 GHz _____ 93
   4.4 Microwave performance of the investigated mixtures _____ 98
5 Summary _____ 101
6 Zusammenfassung _____ 103
7 Literature _____ 105

Introduction

Liquid crystals (LC), sometimes called Mesophases, are characterized by properties both of a conventional liquid and of a conventional solid crystal. It was Friedrich Reinitzer in the year 1888 [Reinitzer88] who described for a first time such mesogenic properties via optical investigations of a certain Cholesterinderivate. Because of the peculiar properties of the LCs as fluids and as solids, at the same time fluidity as a remarkable liquid property as well as anisotropy of electric and magnetic properties, optical and dielectric anisotropy etc. exist. LCs were widely studied in the past decades. Nowadays LCs are pronounced materials for different applications like displays, optical switchers, memory storage applications, stress detectors, polarimetry etc. In order to achieve desired properties of devices, in general not single LC molecules are used; instead multi-component mixtures on the base of different LC compounds must be designed. So far only Nematic LCs (NLC) are in use for commercial applications. Beside nematic LCs several distinguishable LC phases exist. One of them is Ferroelectric LCs (FLC), predicted in 1975 by Meier [Meier75] and later on chemically in form of prototypes synthesized. FLCs are characterized by low response time (in the range of 1-100µs). Therefore they can be considered for applications where the response time of NLCs (in the range of 10-100ms) is too slow.

Because of the commercial applications of LCs, their electrooptical properties were intensively studied, being still a permanent point of interest. There are many discovered and described electrooptical effects as are for NLCs for instance: S, B, twist, supertwist, dual-frequency effects [Blinov94], and for smectics SSFLC (Surface stabilized FLC), DHFLC (Deformed Helix FLC), V-shape [Clark80], [Inui96].

In the recent time there is an increased interest on characterizing LC materials as a tunable passive unit for microwave applications (phase shifters, varactors), instead of the use of typical semiconductor or ferroelectric materials. In the recent years it has been shown that NLCs can be used for these purposes with good success. In [Weil03a] a figure-of-merit (FOM) of 110°/dB at 24 GHz was obtained with low tuning voltage for a phase shifter on the base of NLC. Such high value of FOM is not achievable for typical inorganic materials used in phase shifters at GHz frequencies. However one must admit that the response time of NLCs is in general larger as those of, for example, semiconductors. Another drawback of using NLCs is the requirement of thick layers of the material for the use in the microstrip geometry (more than 25µm). If the thickness would be decreased then the width of the strip line should be increased in order to keep the characteristic impedance at 50 Ohm. However this would lead to the increase of the conducting losses. On the other hand at such thick layers the response time of NLCs can reach the order of 1 second or even more (the response time is proportional to the thickness to the power two [Binov94]), what is of course unacceptable for many applications.

On the base of NLCs different kind of microstrip lines were fabricated and investigated [Dolfi93], [Guerin97], [Kuki02], [Martin03], [Weil02].

FLC materials do not have these drawbacks which are typical for NLCs. The response time is much lower and does not suffer from the increase of the thickness of the layer as NLCs do. However because of the complexity of the work with FLCs and orientations problems at high thicknesses there are almost no data available on the performance of FLCs in the microwave region. In some recent reports like [Fujikake03] the performance of the microwave devices based on FLCs could be estimated, however no extraction of material parameters was possible. For geometries like a coplanar waveguide the properties of FLC mixtures should be extremely well tuned in order to achieve a good performance [Moritake05].

On Table 1 for demonstration, a comparison of typical ferroelectrics films and LCs materials is presented.

Table 1 Comparison of Nematic LCs with Ferroelectric films (Data available in literature) [Mueller04], [Mueller05], [Weil03a], [Penirschke06]

Technology 1…40 GHz	Tunability	FoM [°/dB] @ 24 GHz	Control Voltage	Response Time
Ferroelectrics Thin films Thick films	Moderate to High	30 … 60 < 30 (50)	< 20 V 10–100 V	Fast < 1 ns < 1 ns
Nematic Liquid Crystals	Moderate	<110	< 30 V	Slow > 100 ms
Ferroelectric Liquid Crystals	?	?	< 10-50 V	Moderate >1µs

So far the dielectric parameters of LCs are well known in the range of up to 1GHz. In this frequency range there are several relaxation processes present (Molecular modes, Collective modes) [Haase03]. However there is not much known about properties in the higher frequency region. Few publications about the use of time domain spectroscopy exist, but they are limited in the frequency range, or measurements were done at fixed frequencies [Bose87], [Haase03], [Utsumi04]. So far only one publication reported on the properties of some commercial LC mixtures in the wide frequency range 10-110GHz [Mueller05]. Another problem arises because of the fact that in many publications commercial mixtures with unknown compositions or simple single compounds were used. Therefore it is impossible to extract information about how the different single compounds in mixtures and their chemical structures influences the performance in the microwave region. But this information is needed in order to find the right way to design mixtures with good properties for the use in devices like phase shifters, array antennas etc. This information should also help to explain the origin of dielectric losses in the microwave region.

Aims of this work

1) Investigation and optimization of different kinds of LC compounds and mixtures in the microwave region in order to check the influence of the properties of different classes of compounds.

2) Design and investigation of NLC mixtures with high values of optical anisotropy.

3) Design of FLC mixtures with large values of tilt angle and good quality of orientation at thicknesses above 25µm.

4) Dielectric study about the influence of molecular modes of NLCs, present in the low frequency region, on properties in the microwave region.

5) Design and preparation of mixtures with low losses.

1 Liquid crystals

The Liquid Crystalline state is in scope of many physical and thermodynamic properties a special state between the crystalline and the isotropic ones. Therefore the Liquid Crystals are sometimes called Mesophases. Nowadays a huge amount of species are known forming such Mesophases. They can be characterized by properties which describe usually the liquid state (fluidity, isotropy etc.) and at the same time by properties which describe the solid (crystalline) state (optical and dielectric anisotropies, orientational ordering etc.). In general, single compounds and mixtures which exhibit one or several Mesophases are called Liquid Crystals (LC).

There are two main classes of Liquid Crystals: Thermotropic and Lyotropic ones [de Gennes93]. Thermotropic Liquid Crystals are materials which exhibit one or several Mesophases between defined temperatures. Usually such materials possess a geometrical anisotropy showing a rod like (Calamitic LCs) or a disk like (Discotic LCs) shape. Lyotropic Liquid Crystals are materials showing the properties of a Mesophase in a characteristic concentration range obtained by dissolving (or mixing) a certain material in a solvent (usually water), to a broad extent depending additionally on temperature.

In terms of orientational and positional ordering there are two types of Calamitic Liquid Crystals, being exclusively important for this work: Nematics and Smectics.

1.1 Director and Order Parameter.

The direction of preferable orientation of molecules in liquid crystals is characterized by the vector **n**, which is called director. However the director does not provide information about how uniformly the phase is oriented [Blinov94]. Therefore there is a need to describe the degree of ordering, which is in a simple way described by the order parameter, usually with the symbol S: $S=1/2(3\cos^2\Theta-1)$, where Θ is the angle between long axis of the molecule and the director of the LC. The averaging happens on all molecules in the system over the time. In the ideal case S is equal 1, or with director perpendicular to the plane equal 0.5. In the case of the isotropic state S=0, that means the phase has no order. In the real aligned LC systems S is usually in the range of 0.5-0.9. One should note that S depends on the temperature and decreases with the increase of temperature. The order parameter has the same symmetry as nematic phase; that means that S remains unchanged if molecules will be rotated through an angle of 180°. The Order parameter can be also described in the case of the uniaxial nematic phase in a more complex way as a symmetric second-rank tensor:

$$Q_{\alpha\beta} = S(n_\alpha n_\beta - \frac{1}{3}\delta_{\alpha\beta})$$

where $n_\alpha n_\beta$ is the quadratic representation of the unit vector **n**, S is the scalar order parameter. S parameter can be written in the following form $S = \langle P_2((a_i \cdot b_i)) \rangle$. In this case $\mathbf{a_i}$ and $\mathbf{b_i}$ are unit vectors in the direction parallel to long axis of molecule and perpendicular. P_2 is the second Legendre polynominal. In our case of a uniaxial nematic phase, the distribution function f (describes probability of finding a molecule with a given orientation) depends only on the angle Θ. Therefore S can be presented in the following way:

$$S = \frac{1}{2}\int P_2(\cos\Theta)f(\cos\Theta)d\cos\Theta.$$

In the case of mixtures based on different uniaxial nematic compounds each with own, different order parameter, the resulting order parameter depends on the amount of each compound and the internal organization of the mixture.

The order parameter S can be described as anisotropic part of the tensor of the magnetic susceptibility [Chapter 2.4]; with other words the anisotropic part of the second-rank tensor χ can be used as an order parameter.

1.2 Nematics

Nematic liquid crystals are characterized by long-range orientational order, whereas the long axis of molecules is oriented in a preferable direction [de Jeu88]. However the centers of mass of molecules don't show positional order. The director **n** in nematics is oriented parallel to the long axis of molecules, moreover **n**=-**n** is valid. Nematics have the point symmetry $D_{\infty h}$ and

this symmetry prohibits the existence of a macroscopic dipole moment. Aligned nematic LC act as uniaxial optical system where the optical axis is parallel to **n** (examples for biaxial nematics are under discussion). The phase transition from the isotropic state to the nematic state is of first order, with a change of the order parameter S [Blinov94].

If the nematic phase is created out of optically active molecules, than the director, lying perpendicular to the molecules long axis, will rotate in each (hypothetical) layer, thus a helical structure will be created. Such nematics are called chiral-nematic or cholesteric LCs. The period along the director, were the same (hypothetical) layer arrangement reappeared, namely by turning the helical axis around the director by 2π, is called the helical pitch. The value of the helical pitch depends on the properties of the phase on a molecular level and on the temperature. Cholesteric LC's are uniaxial systems with the optical axis parallel to the helical axis, thus perpendicular to the long axis of molecules.

1.3 Smectics

Smectic LCs exhibit not only orientational order but also some translational order. The molecules are grouped into layers, enforcing positional order in one direction. Inside of the layers the ordering can be described by the nematic order parameter [Chapter 1.2]. There are different types of packing of molecules in the layers; each of these types corresponds to different kinds of smectic phases. The smectic phases are indexed with Latin letters A, B, C etc. The most important smectic phases are SmA and SmC phases. In the smectic A phase, the molecules point perpendicular to the layer planes, whereas in the smectic C phase, the molecules are tilted with respect to the layer planes.

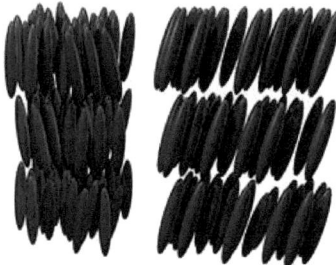

On the above picture one can see the schematic representation of molecules aligned in the smectic phases. The smectic A phase (left) has molecules organized into layers arranged perpendicular to the layer plane. In the smectic C phase (right), the molecules are tilted inside the layers. In the case of SmC and other tilted smectic phases such parameters like order parameter S and density distribution inside the layers are not enough to describe the mesophase like in the SmA or other non-tilted phases. One needs another two parameters in order to describe such phases. These parameters are the tilt angle (angle between layer normal and long axis of molecules) and the azimuthal angle of the director **n** in some fixed coordinate system. If the smectic phase is created by the chiral components (not necessary only) than the SmC* and SmA* phases will be formed. In the case of SmC* phase the effect of chirality will create a helical structure, where the azimuthal angle of the director will continuously change from layer to layer (FLCs).

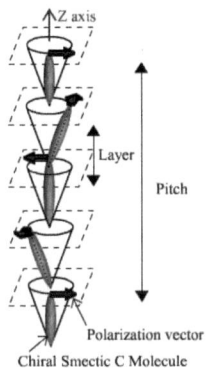

Chiral Smectic C Molecule

1.4 Electrooptical effects in nematic liquid crystals

The majority of electrooptic effects in liquid crystals are based on the following: because of the dielectric (or magnetic) anisotropy, molecules forming the liquid crystals experiences the rotational moment in order to lower the energy if the field is applied. Because of the relative low viscosity of liquid crystals it is possible to reorient molecules during a short time scale (down to several ms for nematics). Because of the high anisotropy of the optical properties of liquid crystals it is possible to reorient the molecules under applied external field. Therefore most of the electrooptic effects in LCs are based on this reorientation. Although the major property is the dielectric or magnetic anisotropy, in addition, the reorientation strongly depends on other properties like viscosity, elastic constants, and the initial orientation of molecules in the volume.

The initial orientation of the director in the LC cell is very important for all electrooptical effects. Hence, the distribution of the director in the cell depends on the properties of the mesophase (nematic, smectic) and on the orientation of molecules due to the orientating layer. There are three types of molecules orientation possible (with respect to the aligning layer): homeotropic (molecules are perpendicular to the plane of the cell), planar (parallel) and tilted. In case of electrooptic effect after the field is removed, there exist a moment of power which tends to return the obtained orientation back to the initial one. The deformation of the LC layer leads to the change of the optical properties. The main effects for nematics are S, B effects and twist effect.

The S-effect (Frederic's transition) is the reorientation of the planar aligned nematic liquid crystal with positive dielectric anisotropy under the applied electric field. The B-effect follows the same principles; however the initial orientation of the molecules near the electrodes is homeotropic. The investigation of nematics follows mainly the use of testing cells based on the S and B effects.

The characteristic response time for S and B effects can be calculated from the following equations

$$\tau_{on} = \frac{4\pi\gamma_1 d^2}{\Delta\varepsilon V^2 - 4\pi^3 K_i} \quad \tau_{off} = \frac{\gamma_1 d^2}{\pi^2 K_i}$$

where d is the thickness of the nematic layer, V is applied voltage, $\Delta\varepsilon$ is dielectric anisotropy, K_i is the elastic constant which corresponds to related electrooptical effect and γ_1 is the rotational viscosity. One should note that response times are proportional to d^2 for nematic LCs.

1.5 Surface Stabilized Ferroelectric Liquid Crystals (SSFLC)

Clark and Lagerwall demonstrated in 1980 for FLCs the possibility to create cells with macroscopic polarization [Clark80]. Such cells are called surface stabilized. Here in the cells the helix is completely unwounded. This can be obtained if the thickness of the cells will be several times smaller than the helical pitch. In this case under the influence of alignment layers the helix will be unwounded and the whole structure of the smectic phase becomes polar. Such system is characterized by two or several thermodynamically stable and optically different states. The switching between these states is due to the applied electrical field. In SSFLC's the molecules are arranged in kind of layers, this structure is called bookshelf geometry. The molecules can move around the cone.

In every SmC* layer the vector of spontaneous polarization is in the plane of the layer and perpendicular to the plane which goes through the normal to the layers and the director **n** of the LC. In order to switch molecules between the SSFLC states one need to apply charges of $2P_s$. The dynamics of switching can be described by the Sine-Gordon equation.

$$\varphi(t) = 2\arctan(\tan\frac{\varphi_0}{2} e^{-t/\tau})$$

φ_0 is the angle between $\mathbf{P_s}$ and **E** at t=0. τ is the response time.

One can see that the response time is inversely proportional to the electrical field. Some of the measurement techniques presented in [Chapter 3.2] were done with the help of prepared SSFLC cells.

2. Structure of molecules and physical properties of the LCs

2.1 Thermostability of the mesophases.

Most of the thermotropic liquid crystals can be described by the formula

$R_1 - (-A-)_n- Y- (-A-X-)_n-R_2$.

In this case A is a cyclic fragment, usually aromatic or heterocyclic. Y is the bridge fragment and $R_{1,2}$ are alkyl chains or some groups (CN, OCH_3, NCS, F etc.).

One of the challenging properties of liquid crystals is their phase transition temperatures, especially the melting and the clearing point. The clearing point Tcp is the temperature at which the liquid crystal exhibits the phase transition between mesophase and isotropic state. At this temperature the kinetic energy of thermal motions of molecules become equal to the energy of the molecular interactions, which produce the far order orientation.

The Maier-Saupe [Maier57] theory pointed out that the main role in the organization of the nematic phase is due to the dispersive attraction energy between the molecules. Therefore molecules with a remarkable anisotropy of the polarisability are needed for the creation of the nematic phase. Increase of this anisotropy leads the increase of the temperatures of the phase transition including the clearing point. Anisotropy of polarisability consists of the core part and tails part,

$Tcp \sim (\Delta\sigma_m)^2 + 2 \Delta\sigma_m \Delta\sigma_{R1R2} + (\Delta\sigma_{R1R2})^2$

$\Delta\sigma_m$ is the polarisability of the core and $\Delta\sigma_{R1R2}$ those of the tails. In case when $\Delta\sigma_m > \Delta\sigma_{R1R2}$, we can neglect the last part and can receive a linear dependence between Tcp and $\Delta\sigma_{R1R2}$. Experimental data shows that this is in good agreement between the Maier-Saupe theory and experiment [Maier57].

The Maier-Saupe theory can not explain all kinds of effects [2.1.3]. There are other models which can be used for the description of the LC phases. One of this is the Onsager hard-rod model [Onsager49]. This theory considers the volume excluded from the center-of-mass of an idealized cylinder. Specifically, if the cylinders are oriented parallel to one another, there is a very little volume that is excluded from the center-of-mass of the approaching cylinder (it can come quite close to the other cylinder). If, however, the cylinders are at some angle to one another, then there is a large volume surrounding the cylinder where the approaching cylinder's center-of-mass cannot enter (due to the hard-rod repulsion between the two idealized objects). Thus, this angular arrangement sees a decrease in the net positional entropy of the approaching cylinder (there are fewer states available).

While parallel arrangements of anisotropic objects lead to a decrease in orientational entropy, there is an increase in positional entropy. Therefore in some cases a greater positional order will be entropically favourable. The Onsager theory predicts that a solution of rod-shaped objects will undergo a phase transition, at sufficient concentration of rods, into a nematic phase. Recently this theory became used to describe the phase transition between nematic and smectic-A phase at very high concentrations [Hanif06]. Although this model is conceptually useful, the mathematical formulation is with several assumptions limiting its applicability to real LCs systems. The main difference between Maier-Saupe and Onsager model is that the first describes Nematic-Isotropic transition in thermotropic LCs and the second the transition at some point where the volume fraction of rod-shaped object is increased.

Another model which can describe LC systems is the elastic continuum theory [Govers84]. The Liquid crystal material is treated as a continuum; molecular details are entirely ignored. This theory considers perturbations on a presumed oriented sample. One can identify three types of distortions that could occur in an oriented sample: (1) twists, where neighbouring molecules are forced to be angled with respect to one another, rather than aligned; (2) splay, where bending occurs perpendicular to the director; and (3) bend, where the distortion is parallel to the director and usually to the main axis of the mesogen. The response of the material can then be decomposed into terms based on the elastic constants corresponding to the three types of distortions. The elastic continuum theory is particularly powerful for modelling liquid crystal devices.

2.1.1 Influence of R_1 and R_2.

The most used groups or fragments at the terminal position of the above presented general scheme of a LC molecule are –CN, -R, -OR, -COOR, -OOCR, -NCS, -OCF$_3$. R is C_nH_{2n+1} or a branch unit; H can be substituted by any other units, as fluoro for example. For most of LC components n takes values in the range of 1-12. All of these fragments influence differently the clearing point. Additionally alkyl chains will change this temperature depending on the length.

For some classes of compounds if we will take the highest n numbers, the Tcp (temperature of the clearing point) will increase. However for high temperatures (>80°C) one can see a different dependence [Grebenkin89]. In many cases the dependence of Tcp from the n number will have alternating sequence, the so called odd-even effect. That means Tcp will increase if we compare LC molecules with even n number with molecules with next odd n number, on the other hand will decrease when we compare with the next molecule with the next even n number etc. This is in a good agreement with the Maier-Saupe theory. By changing from the even n to odd n, we will add CH_2 to the alkyl chain. In this case, taking in account the geometry of the chain, we will increase only the length of the molecule but not the breadth. In the case of changing from odd to even n, both length and breadth will change. Therefore the l/h ratio (l is the length of molecule and h is the height) will be lower and Tcp will be lower. However one should note that this rule is not true for all LC materials.

At the same time the length of the alkyl chain influences the formation of the mesophase. Usually nematic compounds have the n number in the range of 1-5. Groups like CN or NCS at the terminal position usually create the nematic phase.

Not only the length of the chains plays a role but the type of the chain or group at the terminal position have strong influence on the temperatures of phase transitions and especially on the Tcp.

For example for the 4-alkoxy-4'-propylbenzene compounds

$$C_3H_7-\text{[benzene]}-N=N-\text{[benzene]}-OC_nH_{2n+1}$$

where n=3-10, the temperature of clearing point will change in the range between 65-91,5°C [Demus84]. For the same compound but with alkyl chain C_nH_{2n+1} at the terminal position instead of alkoxy the clearing temperature will be in the range of 32-69°C. However the difference in the temperatures of melting and clearing between these groups will depend on the length of the chains. Another example is the well known 5CB compound and its analogue with alkoxy group [Demus84].

$C_5H_{11}-\text{[biphenyl]}-CN$ $\qquad$ $C_5H_{11}O-\text{[biphenyl]}-CN$

Cr 24 N 35 Iso $\qquad\qquad$ Cr 48 N 68 Iso

As we can see the thermostability of the nematic phase is higher for alkoxycyanobiphenyls compared to alkylcyanobiphenyls, however one should note that the temperature of melting also increases. Cyano group at the terminal position usually increase Tcp in comparison with the alkyl chain, but decrease it in comparison with the alkoxy chain. The amount of the change of the phase transition temperature is strongly dependent on the length of the chains. Another possibility to change the properties is the introduction of polar units, like Fluoro, instead of H, to the alkyl chains. In this case such tails produce a very good lamellar packing and therefore it will tend to create smectic phases. Usually such components are characterized with high thermal stability and a broad range of smectic phases because of the decreased melting temperature.

2.1.2 Influence of Core

As was shown previously the temperature of melting and clearing strongly depend on geometrical properties of molecules, as length/breadth ratio. By increasing the number of rings in the core, this geometrical ratio will increase. That leads to a simple conclusion that the temperatures of melting and clearing will also increase. Mixtures on the base of four or five membered rings will hardly reach the LC phase at room temperature. For example quaterphenyl components without lateral substituents or bridge fragments can have Tmp (temperature of melting point) at 200°C or even higher.

The influence of different rings in the core on the temperatures of phase transitions can be compared with 1,4- substituted phenyl rings in the core.

Taking into account the fact that the thermostability of the nematic phase depends on the geometrical anisotropy of molecules we can expect that by the substitution of the phenyl rings with 1,5 or 1,4 naphthyl fragments the clearing point will decrease. If we look at 1,4 naphthyl fragment we can see that this fragment in the molecule will decrease the l/h ratio.

1,4 naphthyl

For example for the following component the clearing temperature is equal 340°C in the case if A is 1,4- phenyl and equal to 282 °C if A is 1,4 naphthyl [Demus84].

CH_3O—〈 〉—CH=N—A—〈 〉—OCH_3

If we substitute the phenyl ring with 2,6 naphthyl fragment, the situation will be different because even if the molecule will be breather, at the same time the length will increase. In this case the temperature of the clearing point will be higher.

Another widely used fragment is the cyclohexane ring. The influence of this fragment on the properties is different and depends on other structure fragments. In some cases we will increase the temperatures of melting and clearing. For example, the substitution of one phenyl ring in 5CB versus cyclohexane (to PCH's) increases Tcp by 21°C. However the temperature of melting is also higher.

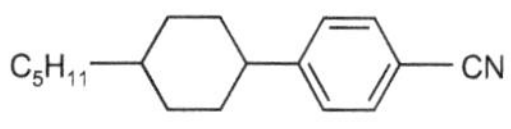

$$C_5H_{11}-\bigcirc-\bigcirc-CN$$

Cr 31 N 55 Iso

By substituting the second phenyl (to CCH's) the Tcp will be even higher and equal to 85°C. One can see that even with a decrease of the polarisability such components still form a mesophase and their thermal stability is even higher; this shows again that the Maier-Saupe theory is not useful for some classes of molecules.

Another conclusion is that the shape of molecules plays an important role in the formation of the mesophase. For example for compounds shown below the thermal stability is higher if A is a phenyl ring (Tcp=281°C), compared to A as cyclohexane ring (Tcp=243°C). In this case the melting point for component with cyclohexane ring will be lower.

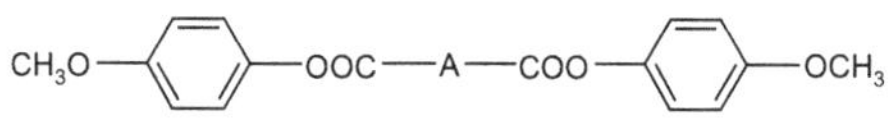

$$CH_3O-\bigcirc-OOC-A-COO-\bigcirc-OCH_3$$

One should note that in normal case the introduction of a cyclohexane instead of phenyl will lead to lower melting temperature. However as was shown there are exclusions.

In comparison to 2 ring compounds the addition of another one or two ring fragments will increase the temperatures of phase transitions because of the increase of anisotropy of polarisability and of the shape of the molecules.

By substitution of the phenyl ring with some heteroaromatic fragment the geometrical anisotropy is mainly not changed. However the presence of hetero atoms in the cyclic fragments can strongly influence the polarisability, the angles between fragments inside the molecule and the interactions between molecules and therefore will influence the temperatures of phase's transitions. One of the examples is the pyrimidine ring

This fragment is mostly used for creating components with smectic phases. Pyrimidine fragment and other cyclic fragments with N atom (like pyridine) have the tendency to decrease the temperature of both, melting and clearing. Therefore phenylpyrimidins are oft used as base matrixes for creating mixtures with a smectic C* phase at room temperature.

2.1.3 Influence of Bridge fragments

Typical bridge fragments used for creating liquid crystal materials are single bond -, azo – N=N-, ester COO, acetylene –C≡C-, imine –CH=N-, ethane –CH$_2$-CH$_2$-, -CH=CH- and others. In general adding of a bridge fragment into the structure tends to destroy the lamellar packing and therefore lead to a nematic phase in a series of compounds which commonly show a smectic phase, however there are a lot of exceptions.

Fragments like COO, widely used for the preparation of liquid crystals, increase the length of the molecule and therefore the polarisability and as consequence the clearing point becomes

increased. Because this bridge is "broken" it also increases the breadth of the molecule. For example introducing the COO fragment into the structure of 5CB will increase the N-Iso transition up to 55°C [Osman81].

The acetylene bridge fragment also increase the anisotropy, because this group will keep the linearity of the molecule and will increase the longitudinal polarisability, therefore the temperature of phases transitions will become higher.

In most cases the influence on the thermal stability of the molecule can be explained by geometrical factors. For bridge fragments like azo group or imine group the length is

increased and so the longitudinal polarisability and as a consequence the thermal stability too. By using such long bridge fragments like -CH=CH-COO- it's possible to increase the temperature of the N-Iso transition even more, however such groups are not very suitable for practical applications. In [de Jeu88] the influence of several different bridge fragments on the properties of this component is described.

There are the following dependences with different X:
COO (41,5°C) < CCl=CH (51,4°C) < CH=N (62,9°C) < N=CH (63,7°C) < N=N (65,4°C) < CH=CH (124,5°C) (in this case monotropic N-SmB transition is also present).
Similar results were obtained by [Titov75] for the series of compounds with R= C_nH_{2n+1}, $C_nH_{2n+1}O$, $C_nH_{2n+1}COO$

R—⟨⟩—X—⟨⟩—CH=CH-CN

This paper support the above presented tendency on the increase of the N-Iso transition in the sequence COO < CH=N < N=N < N=N(O) < CH=CH-COO.
The group CH=C(CN) has the tendency to decrease the clearing point because of the increased breadth and also the polarisability in the perpendicular direction.
For most of the bridge fragments not only Tcp will be increased but also the melting point. Some classes of compounds (for example some biphenyl components) exist were groups like N=N or CH=N will not influence the melting point or the influence will be much lower than the change of the Tcp. Nevertheless, such fragments broaden the range of the mesophase what is also important.
For the following molecule there is a big difference in Tmp

C_2H_5O—⟨⟩—X—⟨⟩—C_4H_9

when X is N=CH (the melting point is 60°C) or X is CH=N (37°C). Similar examples exist for X=COO (91°C) and X=OOC (59,7°C) [Grebenkin89].

2.1.4 Influence of Lateral substituents

As lateral substituents atoms as F, Cl, Br and groups CN or CH_3 are usually used. In terms of geometrical anisotropy of molecules such lateral substituents attached to the cyclic fragments increase the breadth of the molecule, therefore in most cases the temperature of the clearing point is decreased. It is also clear that the size of such substituents plays an important role. Lateral substituents also tend to destroy the lateral attractions between molecules and therefore they are more preferable for creating the nematic phase instead of a smectic one. [Osman85] reported transition temperatures for compounds without and with different lateral substituents.

Without substituent: Cr 50 SmC 196 Iso and with

Y=F	Cr 61 SmC 79,2 N 142,8 Iso
Y=Cl	Cr 46,1 N 96,1 Iso
Y=CH$_3$	Cr 55,5 N 86,5 Iso
Y= Br	Cr 40,5 N 80,8 Iso
Y=CN	Cr 62,8 (SmC 43,1) N 79,5 Iso

The clearing point is decreased in case of Cl, Br and methyl group, the smectic phase was destroyed and the nematic phase was formed. In case of F and CN groups, the smectic phase is still present, may due to the increased lateral attractions. The influence of the lateral substituents is dependent on the position of e.g. the Fluor atom.

For example, for the series of biphenyls Tcp is decreased in the range of (34-41°C) when F is at position marked (1), and in the range of (13-20°C) at the position marked (2), depending on the length of the chain.

At the same time the position of the substituent can influence the phase behaviour of the component. For example for a terphenyl fragment the position of two fluoro substituents strongly influences the phase sequence [Gray89].

Cr 60 N 120 Iso   Cr 81 SmC 115,5 SmA 131,5 N 142 Iso

Other widely used substituents are the methyl group and Cl. In this case they usually don't enhance the thermal stability of the smectic phase and only enhance the nematic phase. Likewise it was shown previously, Cl can also decrease the melting point of the molecule and therefore this type of lateral substituents is often used to reduce Tmp.

The long alkyl chain as lateral substituent can drastically increase the breadth. Therefore by introducing such fragments it is possible to reduce greatly Tcp [Ivashchenko88]. However such kind of substituents does not have a practical interest because the range of the nematic phase is usually strongly decreased and the viscosity of the system is rather high.

2.2 Thermal stability of LC mixtures

In previous chapters the influences of the structure elements on the properties of single compounds and mixtures were presented. In this chapter the influence of single molecules on the parameters of mixtures will be discussed.

For the calculation of the phase diagrams, with other words the weight percentage of components needed to create eutectic mixtures, the ideal mixing rule is used [Demus74, Pohl77].

$-\ln x_i = (\Delta H_i /R)(1/T - 1/T_i)$

where x_i, T_i, ΔH_i are the molar ratio, the temperature of the melting point and the enthalpy for the i'th-component of the mixture. T is the temperature of the melting point of the eutectic mixture. One should note that this formula can provide precise information only for few cases because this formula describes components which can form ideal solutions. In the case of real liquid crystal systems there are usually very strong intermolecular forces. This will lead to the change of the temperatures of phase transitions and also to appearing of induced LC phases [Ivashchenko76]. For example for the solution of two nematic components, one of them is less polar, the other strong, there is the possibility of forming smectic phases. Another point which makes the calculations of temperatures hard is the fact that some components can have several crystal modifications and therefore different values of melting enthalpy must be taken into account. This can lead to the possibility of forming several different phase diagrams of the same mixture.

For real and even some scientific applications usually multi-component mixtures are used. This is due to the fact that usually the properties of one single molecule cannot fulfill the task. Moreover most of LC components have melting temperatures above room temperature.

2.3 Influence of structural elements on the values of the optical birefringence.

Optical characteristics of the liquid crystal molecules and mixtures in a layer or a cell can be changed by applying external electric or magnetic fields. This is what defines the use of liquid crystals for practical applications in most cases. Nematic liquid crystals are uniaxial with two refractive indexes n_o and n_e. The n_o is the ordinary refractive index and this corresponds to the case when the electrical vector of the linear polarized light is perpendicular to the director of liquid crystals. n_e is the extraordinary refractive index with electrical vector of the linear polarized light parallel to the director. This is true for nematics and uniaxial smectics, where the direction of optical axis is defined by the director **n**. We can define the refractive indexes as $n_\perp$ and $n_\parallel$, therefore $n_o = n_\perp$ and $n_e = n_\parallel$. The value of the optical birefringence is defined as

$\Delta n = n_\parallel - n_\perp$. Usually the value of Δn is larger as 0 and sometimes in the range up to 0,8. The optical birefringence of the liquid crystal system depends on the temperature and the wavelength of the light. With the increase of temperature the value of anisotropy decreases because the value of the order parameter S becomes lower. The value of optical birefringence becomes equal to 0 when the temperature is at the clearing point or even higher.

The dispersion of the optical birefringence is usually present below the visible light because of the high absorption of the light in the UV region. For some classes of compounds the dispersion is in the visible light range, as for example for azo and azoxy compounds.

The values of the refractive indexes depend on the deformation of the (electron) polarisability of the molecules along and perpendicular to the long axis [de Jeu88]. Therefore the optical anisotropy depends on the anisotropy of the polarisability. The conjugating bonds strongly influence the anisotropy of the molecule. Molecules with conjugating bonds in a ring increase the values of the Δn, like phenyl, pyrimidine etc. Therefore molecules without such bonds have lower refraction indexes and anisotropy.

The length of the alkyl chains also slightly influences the values of the optical anisotropy. With the increase of the length the anisotropy decreases, however one must note that in this case it's also possible to see the effect of alteration like for the thermal stability [Grebenkin85]. This can be easily explained by the fact that addition of the next methyl group will increase or slightly decrease the polarisability, when this group is added at odd or even position correspondingly, because of zigzag conformation of the alkyl chains [de Jeu88]

The value of optical anisotropy of molecules can be calculated by addition of the anisotropies of the structural elements.

Molecules with bridge fragments must be specifically considered. For instance, carboxylic fragments reduce greatly the conjugation in the molecule and therefore lead to the decrease of anisotropy. In most cases the bridge fragments has the same influence like the carboxylic fragment but for example such fragment like triple carbon bond between phenyl rings, a tolane group, greatly increases the value of the optical birefringence, because of the improved conjugation. The usual values of the increase of birefringence are about 0.1 or more, depending on the kind of other structural components. However there is drawback because of decreased UV stability of the tolane molecule.

The values of optical anisotropy can be influenced also by groups at the terminal position. This influence is usually not very prominent, except for the longitudinal NCS group which increases the birefringence greatly. Different lateral substituents can also change the optical birefringence of molecules. In the case of single atoms or groups like methyl this influence can be neglected.

As was stated before the values of anisotropy of the liquid crystal layer are dependent on the temperature because of the decrease of the order parameter by increasing the temperature. There is a proportional dependence between Δn and order parameter S, in [de Jeu88] this is described as $\Delta n \sim r/2S$, where r is the density. The density of the liquid crystal materials changes not very much with the temperature and therefore the temperature changes of the values of Δn can provide useful information about the temperature dependence of the order parameter.

2.4 Influence of the structural elements on the dielectric properties of liquid crystals.

Dielectric properties of liquid crystals are very important. The polarization of molecules without permanent dipole moments consists of electronic and ionic parts. Molecules with a permanent dipole moment parallel to the long axis show because of the orientational polarization the tendency to align parallel to the electric field. In the case of nematic liquid crystals the permittivity is characterized with $\varepsilon_\parallel$, which is the permittivity in the direction parallel to the long axis of molecules, and $\varepsilon_\perp$, which is the permittivity in the perpendicular direction to the long axis. Therefore the dielectric anisotropy $\Delta\varepsilon$ is equal to $\Delta\varepsilon=\varepsilon_\parallel-\varepsilon_\perp$, the average dielectric permittivity $\varepsilon_{average}$ is equal to the $\varepsilon_{average}=1/3(\varepsilon_\parallel+2\varepsilon_\perp)$.

2.4.1 Liquid Crystal material with low values of dielectric anisotropy.

This group of LC-compounds show a dielectric anisotropy not higher than 2-3, being compounds not containing strong polar groups like CN, NCS etc.

For such molecules we can neglect any effect of dipole-dipole correlation [de Jeu83], which can greatly influence the dielectric properties of molecules. The dielectric anisotropy depends on the temperature because the order parameter S is temperature dependent. For mixtures based on the components with low values of dielectric anisotropy the additive rule will lead to the following formula $\Delta\varepsilon_{mix}=\sum\alpha_i\Delta\varepsilon_i$, where α_i is the weight percentage of the components in this mixture.

The values of anisotropy should be taken by the same temperature which is defined by the reduced temperature $T=(Tcl-T)/Tcl$. In this case the order parameter S at each reduced temperature will be about the same for all compounds.

2.4.2 Liquid Crystal material with high values of dielectric anisotropy.

Molecules with high values of $\Delta\varepsilon$ are very important for practical applications. Usually such molecules have strong polar groups at the terminal position like CN, F, and OCF_3. For such molecules, the dipoles play a dominant contribution to the dielectric parameters and therefore the formulas $\varepsilon_{average} \approx \mu 2T$ and $\Delta\varepsilon \approx \mu 2S/T$ are than not more valid.

If one would calculate $\Delta\varepsilon$ from these expressions, presented above, using the values of dipoles for single molecular fragments one will receive values which are much higher than the experimental data. This is caused by dipole-dipole interactions. For some components the amount of such molecules can be rather high and this will lead to an effective dipole moment which is lower than μ.

Neighbouring polar molecules in such mixtures will try to align themselves antiparallel to each other, in the way that the ends of one molecule align toward the end of another molecule that has opposite partial charge in order to maximize the attractive interaction between them. If averaged over time, the interactions at all are attractive. How strong these interactions are, depends on how strong are as well dipole moments as the angle between molecules and the distance between them. Also a strong dependence on temperature exists, because the random thermal fluctuations in the volume change the local orientation of molecules at some degree. Dipole-dipole interactions are usually much weaker than dipole-ion interactions in LC materials (usually prepared mixtures have some amount of impurities; this is true even for mixtures used for preparation of LC displays).

2.5 Magnetic anisotropy

The magnetic moment **M**, induced by the external field **B**, can be described with the help of the tensor of the magnetic permittivity. An external magnetic field alters the orbital velocity of electrons around their nuclei, thus changing the magnetic dipole moment in the direction opposing the external field. This leads to a magnetic moment **M**, which correlates with the applied field **B**. In the case of the uniaxial nematic phase, this tensor has two components $\chi_\parallel$ and $\chi_\perp$. Therefore the magnetic anisotropy can be described as

$$\Delta\chi = \chi_\parallel - \chi_\perp = 3/2(\chi_\parallel - \chi), \text{ where } \chi = 1/3(\chi_\parallel + 2\chi_\perp)$$

Usually liquid crystals are diamagnetic [de Jeu88]. That means the values of $\chi_\parallel$ and $\chi_\perp$ are small and negative. The susceptibility of atoms is isotropic and therefore the values of susceptibility for molecules are small. However $\Delta\chi$ is positive for calamitic Liquid Crystals based on phenyl group containing compounds; otherwise it can be positive or negative. This can be explained by the presence of the ring currents, formed by delocalized π-electrons of, for example, phenyl rings.

Because the diamagnetism is temperature independent, χ in the nematic phase is equal to χ in the isotropic phase. In order to receive all information needed we must measure one of the components in the nematic phase and χ in the isotropic phase.

The $\Delta\chi$ for the mixture can be calculated using the additive rule including the group anisotropies, because the intermolecular interaction influences the magnetic properties only weakly.

Because of the magnetic anisotropy of molecules, the LC phase can be oriented in an external magnetic field. In order to minimize the free energy, the director will align parallel to the magnetic field, if the anisotropy is positive, and perpendicular to **B**, if the anisotropy is negative. Such kind of orientation of LC molecules was used in this work in order to align the investigated nematic mixtures [Chapters 4.2.1, 4.2.2]. Typical values needed to orient LC molecules in the nematic phase are around 0.5T (Tesla). In the case of FLC compounds the needed values of the field are much higher because of the higher viscosity, around several T.

2.6 Dielectric properties of LCs.

Materials which do not conduct electrical current or have very high values of resistivity are called dielectrics. The definition of the dielectric permittivity or dielectric constant ε usually needs to measure the capacity of an empty and than filled plane condenser with needed material. In this case the dielectric permittivity can be calculated as

$\varepsilon = C/C_{vacuum}$

The dielectric constant is dimensionless.

After filling the condenser with dielectric material a change of the capacitance appeared. The vector of the dielectric displacement **D** can be defined with the help of the second order tensor ε.

D=$\varepsilon_0 \varepsilon$**E**=$\varepsilon_0$**E**+**P**, where **P** is polarization.

In the case of the nematic phase the dielectric permittivity is the tensor of second order with two components, parallel and perpendicular to the director.

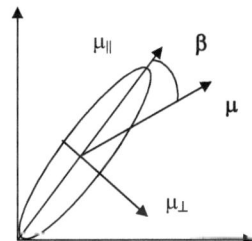

Figure 2.6.1 Components of dipole moments

The dipole moments in these both directions (parallel and perpendicular to the Director **n**) can be described by the following equations

$$\mu_\perp^2 eff = \mu^2 \left[1 + \frac{1}{2}(1 - \cos^2 \beta)S\right]$$

$$\mu_\parallel^2 eff = \mu^2 \left[1 - (1 - \cos^2 \beta)S\right],$$

β is the angle between the dipole moment m and the long axis of molecules, S is the order parameter.

The static dielectric constant can be calculated using the Maier-Maier equations [Maier61]

$$\varepsilon_\parallel - 1 = \frac{NFh}{\varepsilon_0}\left(\overline{\alpha} + \frac{2}{3}\Delta\alpha S + \frac{g_1^\parallel \mu^2 F}{3k_b T}(1-(1-\cos^2\beta)S)\right)$$

$$\varepsilon_\perp - 1 = \frac{NFh}{\varepsilon_0}\left(\overline{\alpha} + \frac{1}{3}\Delta\alpha S + \frac{g_1^\perp \mu^2 F}{3k_b T}(1+\frac{1}{2}(1-3\cos^2\beta)S)\right)$$

Where g is Kirkwood correlation factor, k_b is Boltzmann's constant, F is reaction field, h is cavity field factor, α is polarisability and N is the number density. F and h take in account the field dependent interaction of molecules with the environment. The well-known Kirkwood factor is needed in order to take in account the dipole-dipole interactions. In the case of an isotropic liquid (Order parameter is 0) the Maier-Maier equations transform to the Onsager equations [Onsager]. It is possible to estimate the order parameter of a LC mixture or compound if one will measure the temperature dependence of $\varepsilon_\perp$ and $\varepsilon_\parallel$ if all other parameters are known [Jadzyn99].

The dielectric anisotropy in the case of nematics can be described by the following equation.

$$\Delta\varepsilon = \varepsilon_\perp - \varepsilon_\parallel = \frac{NFh}{\varepsilon_0}\left(\overline{\alpha} - \frac{\mu^2 F}{2k_b T}(1 - 3\cos^2\beta)S\right)$$

As one can see for β values below $\approx 55°$, $\Delta\varepsilon$ is positive, but for values higher than $\approx 55°$ $\Delta\varepsilon$ will be negative.

2.6.1 Dielectric spectroscopy

The dielectric permittivity for liquid crystals is frequency dependent. Therefore the equations for the static constant can not be used in the case if the applied field changes over time. If the field was applied and then is switched off, the orientation polarization will decrease exponentially with a characteristic time τ. The response after changing the orientation of a constant dipole moment by changing the direction of the applied electrical field will take some time. In the case of variable electric fields there is a time delay between the averaged orientation of dipole moments and the applied field. This effect is very noticeable when the frequency is equal or close to $1/\tau$. If the frequency will be much higher, the dipole moment will not be able to orient along the field and therefore there is no contribution of the orientation polarization at such frequencies and only the distortion part plays a role.

In the case of linear dielectric response the induced polarization $\mathbf{P}(t)$ will depend on the applied electric field $\mathbf{E}(t)$.

$\mathbf{P}(t) = \varepsilon_0(\varepsilon^* - 1)\mathbf{E}(t)$

Usually a sinusoidal field is used, thus

$E(t) = E_0 e^{(-i\omega t)}$

ε^* is the complex dielectric permittivity and $\varepsilon^* = \varepsilon' - i\varepsilon''$.

The ε' is also called dispersion component and ε'' the absorption component.

There are two special cases, namely at frequencies 0 and ∞. ε(0) is the static dielectric permittivity and ε(∞) is the value of permittivity at frequencies where only the distortion part plays a role to the polarisability of molecules. The difference between these two values Δε – is the dielectric strength which corresponds to the area below the absorption curve ε''(ω). This value differs from the dielectric anisotropy of liquid crystals, which is also displayed as Δε. The way to study the properties of LC molecules or mixtures is called frequency domain dielectric spectroscopy. There is another approach to study dielectric properties of LC's – time domain spectroscopy (The influence of the sample on the properties of a step pulse propagating through the system is studied) [Haase03]. Experimentally for the liquid crystals there are in general two absorption peaks detectable which correspond to the molecular modes of molecules. Usually such processes in liquid crystals can be described by the Debye equations [Debye45].

ε* = ε' - iε'' = ε(∞) + (ε(0) - ε(∞))/(1+iωτ),

where ω is the frequency of the applied field and τ is the relaxation time, which can be expressed with the help of the critical frequency (f) of the process τ=1/2πf. Imaginary and real part of dielectric permittivity can be defined by the following equations.

ε'(ω) = ε(∞) + (ε(0) - ε(∞))/(1+ω²τ²)

ε''(ω) = ωτ (ε(0) - ε(∞))/(1+ω²τ²)

The maximum value of the imaginary part corresponds to the point of changing the sign of (ε'(ω))' and it's value can be calculated with the help of

ε''$_{max}$ = (ε$_0$ - ε$_∞$)/2.

For the complex dielectric anisotropy so called Kramers-Kronig relationship can be applied, that means that it's possible to calculate the complex dielectric permittivity if ε'' or ε' is known [Kronig26].

From the Onsager model [Daniel67] one can define the dielectric strength with the help of macroscopic parameters.

$$\Delta\varepsilon = \frac{3\varepsilon(0)}{2\varepsilon(0)+\varepsilon(\infty)} \cdot \left(\frac{\varepsilon(0)+2}{3}\right)^2 \frac{4\pi N \mu^2}{3 k_B T}$$

On Fig. 2.6.2 the ε' and ε'' behaviour over the frequency in terms of classical Debye model is presented.

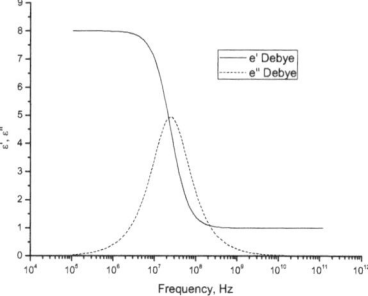

Figure. 2.6.2 Debye representation of the dispersion and absorption parts of some process

Another way to represent the dielectric spectrum is the so called Cole-Cole plot. This plot presents ε'' versus ε'. In the case of a Cole-Cole plot each point of the curve corresponds to the complex dielectric permittivity at one frequency. In the case of single relaxation processes (Debye) the plot looks like a semicircle (if both axes have the same units).
On Fig. 2.6.3 one sees a typical example for a Cole-Cole plot.

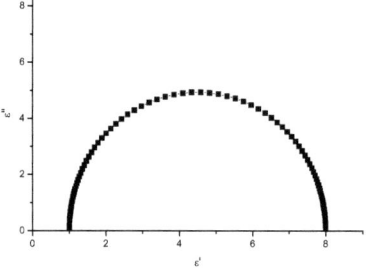

Figure 2.6.3 Cole-Cole plot for the process described in the Fig. 2.6.2

The Debye process can be written in terms of the loss tangent [Haase03].

$$\tan \delta(\omega) = \frac{\omega\tau(\varepsilon(0) - \varepsilon(\infty))}{\varepsilon(0) + \varepsilon(\infty)(\omega\tau)^2}$$

Usually the position of the maximum of the loss tangent does not match the position of the ε'' maximum. The maximum of the loss tangent is shifted to higher frequencies by the factor of the square root of $\varepsilon(0)/\varepsilon(\infty)$. This is sometimes important to overcome some experimental limitations at lower frequencies, for example by FLC mixtures.

The Debye-model describes single relaxation processes only, but many relaxation processes are multifunctional ones. Therefore several formulas are in use in order to simulate the experimental data, among them is the Cole-Cole equation [Cole41] which can be considered after the Debye equation as the most practical ones. The Cole-Cole function describes the relaxation processes in case of broadening the absorption curve which stems from the distribution of the relaxation frequencies. The Cole-Cole formula can be written in the form

$$\varepsilon^* = \varepsilon_\infty + \frac{\varepsilon_0 - \varepsilon_\infty}{1 + (i\omega\tau)^{1-\alpha}}$$

α is the distribution parameter. For $\alpha=0$ this formula will transform into the Debye equation. If α is not zero there will be symmetric spreading of ε''. The half width of the absorption curve will be decreased and the maximum of ε'' is lowered. The origin of the semicircle is moved away from the line with $\varepsilon''= 0$.

In terms of the Cole-Cole formula the real and imaginary part of the dielectric permittivity receive the following form.

$$\varepsilon'(\omega) = \varepsilon_\infty + (\varepsilon_0 - \varepsilon_\infty)\frac{1 + (\omega\tau)^{1-\alpha}\sin(\alpha\pi/2)}{1 + 2(\omega\tau)^{1-\alpha}\sin(\alpha\pi/2) + (\omega\tau)^{2(1-\alpha)}}$$

$$\varepsilon''(\omega) = (\varepsilon_0 - \varepsilon_\infty)\frac{(\omega\tau)^{1-\alpha}\sin(\alpha\pi/2)}{1 + 2(\omega\tau)^{1-\alpha}\sin(\alpha\pi/2) + (\omega\tau)^{2(1-\alpha)}}$$

2.6.2 Dielectric modes in nematic LC systems

Molecular modes are reorientations around the axes of the molecules. Such modes can be usually described with a Debye model with high enough precision.

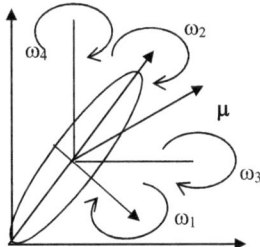

Figure 2.6.4 Molecular modes of nematics

ω_1 is the reorientation around the long axis, ω_2 the reorientation around the short axis, and ω_3 and ω_4 are some precessional motions.

ω_2 for liquid crystals is usually several orders higher than ω_1, which has values close to $\omega_{isotropic}$. For these modes one can write $\omega_1<\omega_3\cong\omega_4<\omega_2$. However ω_4 and ω_3 modes are usually not separable in the dielectric frequency spectrum.

The Kirkwood parameter can be obtained via $g_{\|,\perp}= \tau_{1,2}/\tau_{isotropic}$, $g_\| >1$ and $g_\perp \leq 1$. The $g_\perp$ parameter is acting like the acceleration. As a consequence, the orientational far range ordering makes the rotation around the short axis harder and vice versa around the long axis of molecules easier. One should note that with the increase of the ordering in LC systems the parameter $g_\perp$ becomes lower and $g_\|$ - higher.

The temperature dependence of the relaxation time can be described by the Arrhenius formula
$$\tau=\tau_0\varepsilon^{[Ea/(kT)]}$$
where E_a is the energy of activation of the relaxation process.

3. Experimental part

In this part experimental techniques of measurements of electro-optical parameters of nematic and smectic LC's will be discussed. Also we will talk about X-Ray diffraction measurements of SmA* and SmC* phases. The preparation of LC cells for measurements will be described. An introduction to the principles of measurements of LC materials and mixtures in the microwave region will be given.

3.1 Electro-optical setup

For the measurements of the basic parameters of nematic and smectic liquid crystals and the characterization of the prepared cells our special designed optical setup was used. This allows to measure different parameters of LC's. Simultaneous measurements are possible (see Fig. 3.1.1).

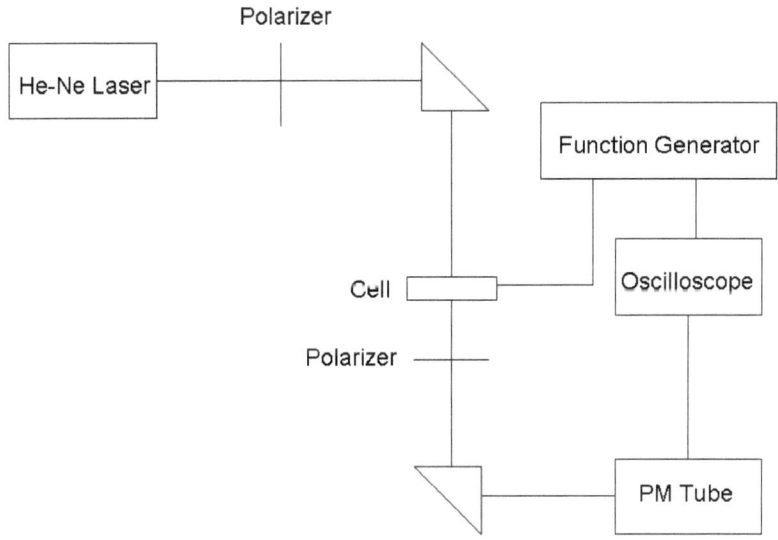

Figure 3.1.1 Experimental electrooptical setup

The He-Ne laser with 633nm is the light source in this setup, his power is 1 mW. To produce linear polarized light, a linear polarizer was installed in front of the laser.

The installed prism directs linearly polarized light into the heating chamber containing the LC cell.

The temperature is controlled by a Eurotherm (Model 905) device. The key part of the setup is the rotation table which is controlled by a custom setup aided by a computer with software written in LabView (v. 6.0).

The rotation is possible in clockwise and counter clockwise directions and implemented by rotation steps of 0.5 degree. The laser beam passed through the cell is directed through the second polarizer (crossed with the first one) to another prism and then to the photomultiplier tube (PM) (Hamamatsu Type R6095). The signal from PM is directed to the oscilloscope HP Infinion. The intensity of the laser beam directed to the PM is controlled by additional filters. The intensity at which the PM tube is working in the linear mode can so be evaluated. LC cells are driven by a function generator (HP 33120A).

This electro-optical setup allows the measuring of the current flow through the LC cell and thus the spontaneous polarization.

3.1.1 Switching time measurements.

To measure the response time of FLC's a square signal is applied to the cell. The applied voltage should be high enough to achieve the full switching of FLC molecules (5-10V per µm). The frequency of the applied signal is usually in the range of 1-10Hz. The orientation of the FLC cell should correspond to the maximum of the contrast, one of the extreme positions of the tilt cone should concur with the orientation of the polarizer or analyzer.

The switching time is defined as the difference in time between two states, which correspond to 10% and 90% of the intensity of the transmitted light.

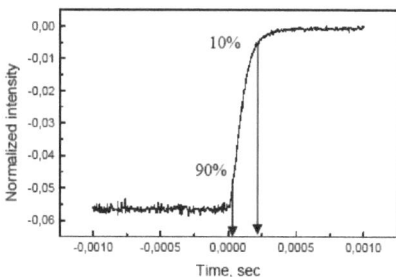

Figure 3.1.2 Electrooptical response of FLC

3.1.2 Spontaneous polarization measurements

Spontaneous polarization of FLC materials can be measured by analysing the repolarization current which flow through the cell. An additional resistor (R = 10-100kOm) is placed in the circuit between photomultiplier and oscilloscope. The current will be transformed into voltage on this resistor. One can see the signal of the repolarization current on Fig. 3.1.3. The response consists of three parts: ohm, capacitor, ferroelectricity (spontaneous polarization). The peak on the graph corresponds to the spontaneous polarization of the FLC material. The value of spontaneous polarization can be calculated using the formula

$$Ps = \frac{S}{2RA}$$

S, the area of the peak, can be integrated using computer software. A is the area of the electrodes in the LC cell. For spontaneous polarization measurements the triangular signal is applied by a function generator.

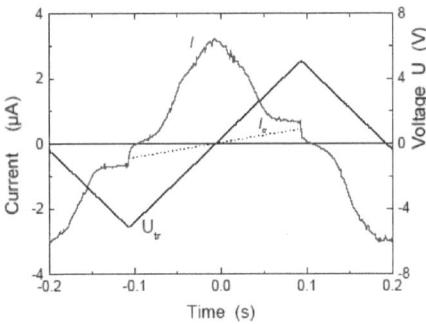

Figure 3.1.3 Measurements of the spontaneous polarization

3.1.3 Tilt angle measurements

The measurement of the tilt angle of the FLC's can be carried out using the presented electrooptical setup under application of the triangular waveform voltage. They are carried out by applying a signal with frequency 1-10Hz and with voltage 5-10V per μm. To measure the tilt angle, the initial position has to be chosen. This should be such a position at which one of the extreme positions of the cone formed by the rotation of molecules in the smectic phase lies in the same direction as the transmission axis of the analyzer. The transmission of the light becomes equal to minimum. By applying the electric field, the transmitted power will increase. By rotating the sample to the position where the electrooptical response changes its sign, one can rotate the transmission axis of the analyzer parallel to the second extreme position. The tilt angle is indicated by θ. By doing such experiment we can measure the value of the angle 2θ.

3.1.4 Polarizing microscopy

For the investigations of phases and temperatures of phase transitions of single compounds and mixtures used in this work, the polarizing microscopy was used.

The samples are prepared by placing the LC's in the isotropic phase onto the object glass, to be covered with a thin glass and than placed in the chamber of the heating table (Mettler) for investigating under a polarizing microscope between crossed polarizers. The temperature was controlled with an accuracy of 0.05°C. The kind of phases and the temperatures of phase transitions can be monitored by controlling the change of the textures of the LC's. In Fig. 3.1.4 (a) one can see the typical nematic texture, so called thread-like texture, formed by threads, which correspond to point defects of creating loops. The typical focal conic texture of the chiral SmA and SmC phases are shown on Fig. 3.1.4 (b and c). The texture of the SmA phase look smooth. By cooling the sample, lines coming in, called strips, indicating the transition to the SmC* phase with a partially developed SmC* helix [Dierking03].

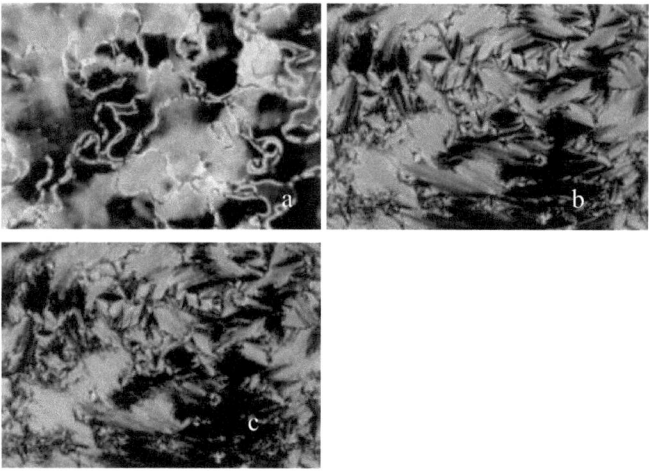

Figure 3.1.4 Textures of liquid crystals a) nematic thread-like b) and c) fan shaped textures of the SmA and SmC phases, correspondingly

3.1.5 Preparation of LC cells

The main tool for characterising liquid crystals is the LC cell. The electro-optical properties of LC's depend on the quality (uniformity of thickness over the cell, alignment material etc.) and parameters of the cells. For this work mainly self-made cells were used. Such cells allow to change parameters and to design a cell for specific purposes. One can vary components of the cell like cell gap by using spacers with different thicknesses. The thickness and the type of the orienting layer can be changed. By using ITO glasses with an additional dielectric layer like SiOx one can protect cells from shortcuts when working with high voltages. Fig. 3.1.5 shows the typical scheme of a LC's cell.

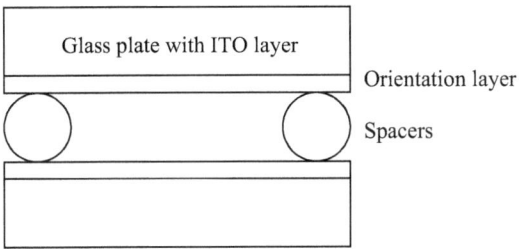

Figure 3.1.5 LC cell.

The cell consists of two glass plates, covered with transparent electrodes (ITO), two orientation layers and spacers, which define the thickness of the cell. The preparation of the cells implemented several technological steps. We used glasses with ITO layers (a gift of Merck), with and without additional dielectric layers. The form of electrodes and contacts is defined by wet chemical etching in HCl solution (12%) with Zn powder. At first the surface of the glass is covered by a special tape which has the desired form of electrodes and is resistant to the etching solution. Than the samples are putted in the etching solution for several minutes, during this step the glasses are additional rubbed. The next step was cleaning the surface of the glass plates with solutions which will remove organic particles from the surface. This needs several steps with the help of heated acetone and hexane. The cleaned surface of the glasses is covered by orienting layers, in our case commercial polyimide and self prepared Nylon-6 solution in 3chlor-ethanol, and then spin coated on. The spin coating process at 3000 rpm produces about 50 nm thick homogeneous polymer layers. The next step is baking the polymer layer in a heating chamber up to 100°C during 30 minutes, holding this temperature for about 1 hour, heating up again to 180°C during 30 minutes, baking at this temperature for 1 hour, and cooling down to room temperature during 2 hours. For the further orientation of LC molecules, the orientation layers were rubbed applying a special technique. Glass or polymer spacers with different diameter are used to provide the needed thickness of the cell.

Here the first technique of controlling the thickness is based on mixing the spacers in some solution, in our case ethanol. This solution is placed onto the cell, close to the edges. After evaporation of ethanol some distributions of spacers over the covered area result. Then, the glass plate is covered by another plate without spacers, using a special mechanical fixing method for both plates. Thermal glue is used to seal the cell.

In another method the spacers are mixed with optical glue, the solution covers the area over the perimeter of the substrates. After putting the substrates together, an UV chamber was used to polymerize the solution of glue and spacers. After this, the cell is sealed by UV glue. This technique provides overall better parameters as the former one. However because of the UV source, the method is not really suitable when thick glasses are used for the preparation. In this case UV sources with high power are needed.

3.1.6 Thickness of the cells

The thickness of the prepared cell can be measured by knowing the area of the electrode A and the empty capacitance of the cell C_0

$$C_0 = \varepsilon_0 \varepsilon \frac{A}{d}$$

Another method of measuring the thickness is based on the interference. By illuminating the empty cell and detecting the outgoing light with a spectrometer supposing the thickness and the refractive indexes of the glass and the orienting layer are known, one can calculate the thickness of the cell. This method is more complicated as the first one but the accuracy is much higher $\sim 0,1 \mu m$.

The filling of the LC into the cell is due to capillary forces.

The ITO electrodes had square resistivity in the range of 20-400 Ohm. Because of the resonance frequency, defined via $f=1/(RC)$, such cells are not suitable for measurements in the high frequency range over 100-1000 kHz because of the so called "ITO-mode". Therefore cells with gold electrodes were used. The square resistivity of gold layers used is in the range of 1-20 Ohm. Experiments with empty gold cells show RC resonance above 10MHz in the case of 15μm thick spacers, the increase of the thickness will shift this resonance to higher frequencies.

Here it should be mentioned that gold cells are not suitable for electro-optical measurements.

For some dielectric measurements of nematic mixtures commercial cells from Poland (Warsaw) with a thickness in the range of 8 μm were used.

3.2 Geometry of microwave measurements
3.2.1 Nematics

Nematic liquid crystals were characterized in the microwave region by using the cavity perturbation method (the setup and calculations were designed in the group of Prof. Jakoby by F. Gölden and Dr. S. Müller) [Penirschke06]. A cavity tube is used. The parameters of this tube are designed related to the resonance frequency. By filling the tube with LC we change the permittivity of the sample and therefore shift the resonance frequency. Measurements of this frequency shift and of the phase shift allow the extraction of the special LC parameters. The cavity tube is filled with the non oriented liquid crystal (diameter of the tube is 500μm, no orienting layers). Under external magnetic field one can orient the LC molecules. If we change **B** into perpendicular direction, we will orient molecules perpendicular to the previous state, see Figure 3.1.6.

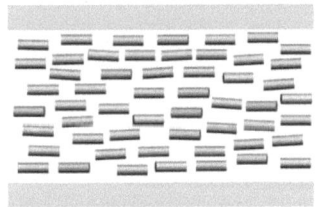

Figure 3.1.6 Planar and homeotropic orientations of LC molecules after applying magnetic field in parallel and perpendicular direction.

3.2.2 Ferroelectric Liquid Crystals

In the case of FLCs a different sample preparation is needed. The idea of FLC measurements is based on the possibility to change the structure from the helical state to the surface stabilized states by applying external voltage, see Fig. 3.1.7. The **E** vector of the microwave passing through the sample is directed perpendicular to the substrates. In this case the permittivity of the system is different in case of applied voltage and without voltage. When the voltage is applied, the **E** vector will be perpendicular to the long axis of molecules. Hence some angle between the long axis of molecules and the **E** vector of the wave result. This angle is dependent on the average tilt angle of molecules. By increasing the tilt angle we will decrease the permittivity because the perpendicular part of the permittivity will be more prominent, therefore the dielectric anisotropy will increase.

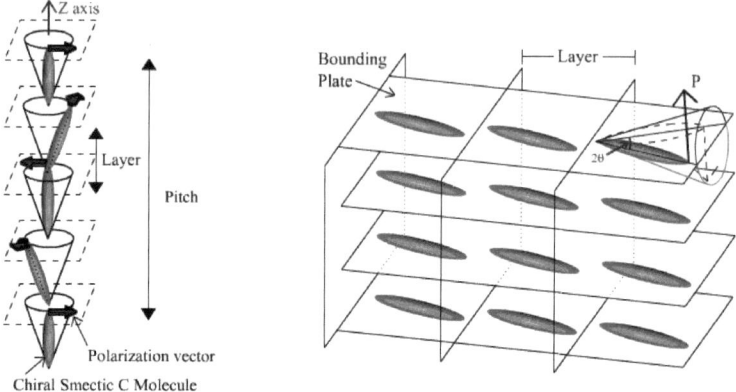

Figure 3.1.7 Helical and SSFLC orientations of molecules without and with applied electric field.

By calculating and extracting material parameters during microwave measurements, the figure of merit (FOM) and tunability (η) of the material are important. These parameters are defined as:

$$\tau_\varepsilon = \frac{\varepsilon_{r,\parallel} - \varepsilon_{r,\perp}}{\varepsilon_{r,\parallel}} \qquad \eta_\varepsilon = \frac{\tau_\varepsilon}{\tan\delta_{\varepsilon,\max}}$$

3.3 Dielectric measurements

Dielectric properties of LC compounds and mixtures were investigated with help of the dielectric bridge (HP 4192A). The measurement range is spread over 5Hz-13MHz. The testing cells are placed in the special constructed heat chamber, and thus the measurements of temperature are possible via controlling by a Eurotherm (Model 818) device. The accuracy of the temperature stabilization is 0.1°C. The measurements of the empty cells and the cells filled with some well know materials like Toluol provides information about the parasitic capacitance of the testing cell. Measurements using the same cells as filled with LC provide parameters like resistance, inductivity and capacitance. These values allow calculating the complex dielectric permittivity ε^*. The measurements and control of the impedance analyzer are forced with the help of computer software. The calculation of the real and the imaginary part of the permittivity is done via a program created in Origin (v.7.0) software. For the determination of the dielectric anisotropy $\Delta\varepsilon$ at 1 kHz a custom made ITO cell and commercial cells were used. The investigated mixtures were measured in both, cells with homeotropic and with planar, orientation. Special designed cells with gold electrodes were used for high frequency measurements. The cell gap of such cells was defined by Mylar foils with different thicknesses (15-20µm).

3.4 X-Ray measurements

The intensity measurements in dependence of the scattering angle 2theta on bulk samples were done using STOE diffractometer and STOE detector (STOE Stadi4). The thickness of a smectic layer or the correlation length of the nematic phase could be obtained. For measurements, LC mixtures or single compounds were filled in special capillaries (Lindemann), which are suitable for such experiments (the thickness of the walls is 0,01µm, the material of the capillaries has no or very low diffraction). After filling, the capillary is placed in a special chamber, where temperature control is possible. The measurements are done on bulk samples without orientation via external fields. The monochromatic x-ray beam is directed onto the sample. The detector is rotating around the sample assembling the intensity at special diffraction angles.

The detector can be rotated from -40 to 40 degrees. CuKα is used as X-ray source. The accuracy of rotating the sample table is 0.001°. The accuracy of the STOE detector is 0.01°. The temperature is controlled by Eurotherm (Model 905) with an accuracy of 0.5°C.

3.5 Investigated mixtures, preparation

For the investigation of the properties of LC materials in scope of this work one need to prepare mixtures based on different compounds. Single compounds are usually characterized by a narrow LC temperature range, higher melting points, mostly above room temperature, and specific physical properties. In comparison, well prepared (eutectic) mixtures perform the broadening of the range of the nematic phase while holding the melting point below room temperature. Depending on the specific requirements the physical properties can be adjusted by selecting the specific components of the mixture.

One specific requirement for experiments in the microwave regime is the increased birefringence.

One way to earn knowledge about materials usable for our purpose is to investigate the properties of some compounds with well suited functional groups like for example

a.) p-thiorhodanid (NCS) groups bonded to phenyl-, biphenyl- etc. units,

b.) p-Cyano-(CN-) biphenyl- or p-Cyano-oligophenyl groups,

c.) carbon-carbon triple (C≡C) bonds, so called Tolane groups

and use those as 'base matrix' by preparing mixtures out of them with different specific components. The last one we will call 'additives'.

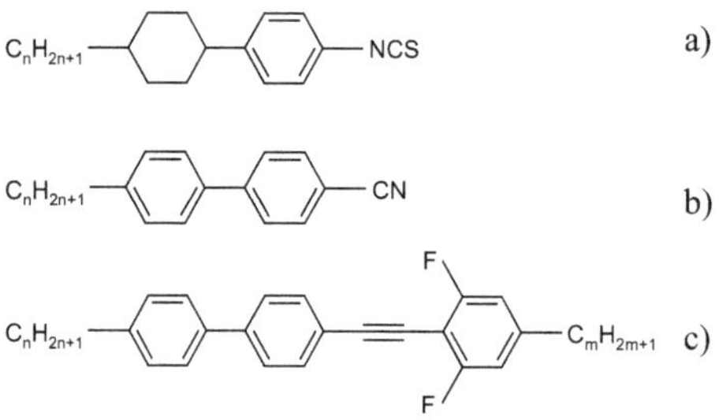

C_nH_{2n+1} —⬡—⬡— NCS   a)

C_nH_{2n+1} —⬡—⬡— CN   b)

C_nH_{2n+1} —⬡—⬡—≡—⬡(F,F)— C_mH_{2m+1} c)

Fig. 3.5.1 Structure formulas of the compounds in the base matrixes. a) NCS compounds b) cyano-biphenyl c) Tolane

The three selected 'base matrix' are presented in Fig. 3.5.1. For the group a), the NCS-containing compounds, we used p-alkyl-cyclohexyl-phenyl-thiorhodanide compounds with different alkyl chains length. The NCS compounds show low viscosity which leads to the decrease of the switching time. NCS containing compounds are well known for their high birefringence up to 0,45.

For the group b), the p-Cyano- containing compounds, we selected the well known 4-Cyano-4'-pentyl-biphenyl (5CB). The properties of this molecule were widely investigated; it shows a nematic phase at room temperature. Some investigations for 5CB in the microwave region were already published [Mueller05], yet their birefringence is about 0,18, what is not high. For the group c), the tolane compounds, we used three different examples all with two lateral fluoro groups, to create a eutectic mixture that will have nematic phase at room temperature. The birefringence of the tolane 'base matrix' is with ca. 0,34 comparable high. Moreover, the range of the nematic phase of the investigated tolanes is comparable broad with a nematic phase up to 140°C. The drawback of the tolane 'base matrix' is the limited UV stability, however this is not so important for experiments in the microwave region where samples and devices are sealed and therefore protected from the day light.

Each of three 'base matrixes' was mixed with one or more single chemical compounds ('additives') in order to improve the properties. In this way we could extract specific properties of such 'additives' in a specific 'base matrix' by knowing the properties of the 'base matrix'.

The general strategy to improve the parameters of LCs is to use additives with high Δn, because the birefringence plays an important role for LCs in the microwave region. Compounds with a long conjugated ring system, e.g. tolane compounds and molecules with the NCS group, fulfil this condition. Within the cooperation with the group from Belarusian State University (Dr. Bezborodov and Dr. Lapanik) terphenyl and quaterphenyl compounds with laterally attached groups were synthesized [Bezborodov06] and investigated. These compounds are with a high value of birefringence up to 0,54.

Additional components with various electrooptical and dielectric properties like terphenyls, compounds with carboxylic and ethane bridge fragments, which broke the rigidity of molecules, molecules with negative dielectric anisotropy, nonpolar components and compounds with different groups at the terminal position were synthesized by Dr. Bezborodov, and investigated and used for preparation of our mixtures in this work.

3.5.1 Additives

There is only a limited number of investigations reported on nematics in the microwave region. Therefore it's needed and important to find out how different classes of single compounds and their chemical structure influence the performance in the microwave region. In this work three different classes of compounds, already introduced, were chosen for creating base matrixes for preparing the mixtures. In this chapter single additives used will be presented and their parameters will be described.

Additive 1 (Add1)

C_5H_{11} — [boroxane] — B — [phenyl-F] — COO — [phenyl-F] — CN

There is a strong longitudinal dipole moment (CN group) and two lateral dipole moments, formed by two Fluor atoms, attached to the phenyl benzoate group. The birefringence and the dielectric anisotropy at room temperature are $\Delta n = 0,17$, $\Delta \varepsilon = 26$. The boroxane ring in the structure increases the dielectric anisotropy. The Fluor atoms as lateral substituents are responsible for decreasing the melting point of this component compared to a compound without lateral Fluor. The phenyl benzoate group restrict the rigid-rod form of the molecule.

[phenyl benzoate structure]

Additive 2 (Add2)

C_5H_{11} — [boroxane] — B — [phenyl-F] — COO — [phenyl-F] — F

This additive has almost the same structure as previous. However the dipole moment in the longitudinal direction is weaker because of substitution of a CN group via Fluor. $\Delta n = 0,165$, and $\Delta \varepsilon = 19$ at Room Temperature (RT).

Additive 3 (Add3)

C₄H₉—⟨cyclohexyl⟩—⟨phenyl⟩—CH₂CH₂—⟨cyclohexyl⟩—⟨phenyl⟩—NCS

This four ring components has an ethane group as bridging fragment. Because of this ethane group there is no conjugation between external fragments possible. There is a strong dipole moment formed by the NCS group in the longitudinal direction. $\Delta n=0,16$, $\Delta\varepsilon=7$ at RT. This compound increases the viscosity. Comparable with the carboxylic fragment, the ethane fragment restrict the rigidity of the molecule.

Additive 4 (Add4)

C₃H₇—⟨cyclohexyl⟩—CH₂CH₂—⟨phenyl⟩—⟨phenyl⟩—NCS

This compound was chosen as alternative to Add3 because of the smaller length of this molecule. It contains a dipolar NCS group and a conjugated biphenyl group. The low frequency diffusive molecular mode should be at lower frequencies compared to Add3 because of the decreased length of the molecule. $\Delta n=0,3$, $\Delta\varepsilon=7$ at RT.

Additive 5 (Add5)

C₅H₁₁—⟨cyclohexyl⟩—CH₂CH₂—⟨phenyl⟩—⟨phenyl⟩—NCS

This additive has increased length of alkyl chains compared to Add4 which leads to small difference between temperatures of melting and clearing. $\Delta n=0,3$, $\Delta\varepsilon=7$ at RT.

Additive 6 (Add6)

C₃H₇—⟨cyclohexyl⟩—⟨phenyl⟩—⟨phenyl(F,F)⟩—OCH₃

This compound has only a weak dipole moment in the longitudinal direction, but two strong dipoles at lateral position, formed by two Fluor atoms. Therefore this compound shows a negative dielectric anisotropy. $\Delta n=0,17$, $\Delta\varepsilon=-6$ at RT.

Additive 7 (Add7)

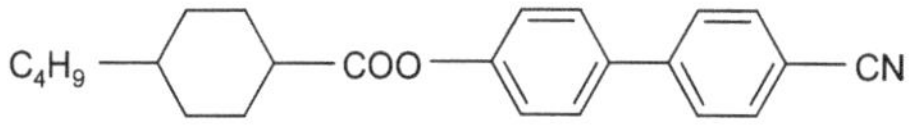

C_4H_9 —⟨cyclohexyl⟩— COO —⟨phenyl⟩—⟨phenyl⟩— CN

This three ring compound is with a longitudinal dipole moment (CN group) and a carboxylic group fragment. $\Delta n=0,30$, $\Delta\varepsilon=11$ at RT.

Additive 8 (Add8)

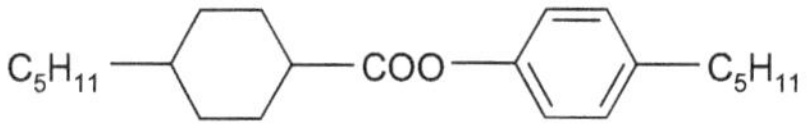

C_5H_{11} —⟨cyclohexyl⟩— COO —⟨phenyl⟩— C_5H_{11}

This additive is with a very weak, mainly perpendicular, dipole moment formed by the carboxylic group. $\Delta n=0,07$, $\Delta\varepsilon=-0,5$ at RT.

Additive 9 (Add9)

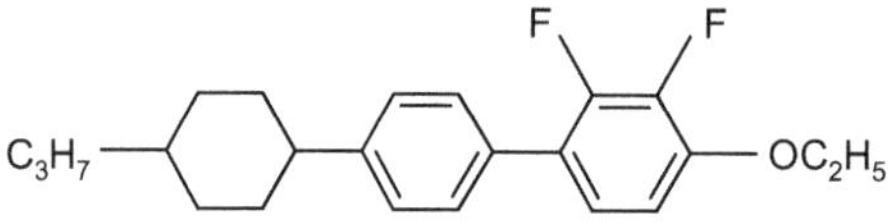

C_3H_7 —⟨cyclohexyl⟩—⟨phenyl⟩—⟨phenyl(F,F)⟩— OC_2H_5

Add9 is comparable to Add6 but with different lateral groups.

Additive 10 (Add10)

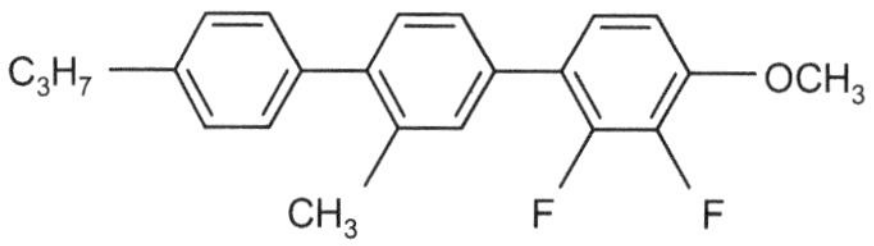

C_3H_7 —⟨phenyl⟩—⟨phenyl(CH_3, F)⟩—⟨phenyl(F)⟩— OCH_3

The compound is with negative dielectric anisotropy; at lateral position exists a methoxy group. This substituted terphenyl molecule is a three phenyl ring-conjugated system. $\Delta n=0,28$, $\Delta\varepsilon=-7$ at RT.

Additive 11 (Add11)

[Structure: C$_3$H$_7$–phenyl–phenyl(CH$_3$)–phenyl–phenyl(F)–F, quaterphenyl with lateral CH$_3$ and F substituents]

Quaterphenyl compound Add11 has a very high birefringence. There are two perpendicular dipolar groups and a longitudinal one. Because of the structure the viscosity is high. In general such compounds show higher melting temperatures (see chapter 2) but because of lateral substituent the melting and the clearing temperatures are lowered. Δn=0,385, Δε=3,6 at RT.

Additive 12 (Add12)

[Structure: C$_3$H$_7$–phenyl–phenyl(CH$_3$)–phenyl–F, terphenyl with lateral CH$_3$]

This terphenyl compound is comparable with Add11, but with some structural change. Δn=0,29, Δε=4 at RT.

Additive 13 (Add13)

[Structure: C$_5$H$_{11}$–boroxane ring–phenyl(F)–COO–phenyl–phenyl(CH$_3$)–C$_5$H$_{11}$]

This four ring compound is a rather complex molecule with a boroxane ring, a carboxylic group and lateral substituents. There exist no strong longitudinal dipolar group but one (Fluoro) at the terminal position. Because of the boroxane ring the resulting birefringence is positive. The viscosity of this compound should be very high. Δn=0,19, Δε=14 at RT.

Additive 14 (Add14)

This boroxane containing compound has a rather unusual structure. The rigidity of the molecule is strongly broken and the l/h ratio is rather small for this four ring compound. $\Delta n=0,21$, $\Delta\varepsilon=13$ at RT.

Additive 15 (Add15)

There is one Fluoro group in lateral substituent. The dielectric properties are a result of the interplay between the strong polar CN group, the boroxane ring and the fluoro group. The dielectric anisotropy is positive and rather high. $\Delta n=0,12$, $\Delta\varepsilon=28$ at RT.

Additive 16 (Add16)

Add16 is a compound with strong lateral dipole moment and therefore it has negative dielectric anisotropy. $\Delta n=0,28$, $\Delta\varepsilon=-2$ at RT.

Additive 17 (Add17)

$\Delta n=0,12$, $\Delta\varepsilon=13$ at RT.

Additive 18 (Add18)

C$_5$H$_{11}$—〈phenyl〉—〈phenyl〉—CN

This additive is 5CB, one of the "matrixes" that was used for the preparation of mixtures.

Additive 19 (Add19)

C$_5$H$_{11}$—[dioxaborinane B]—〈phenyl-F〉—COO—〈phenyl-F,F〉

This is compound is to compare with Add14. It is stronger fluorinated. $\Delta n=0,15$, $\Delta\varepsilon=7$ at RT.

3.5.2 Base matrixes

3.5.2.1 NCS Matrix (BMW1)

For this matrix we used p-alkyl-cyclohexyl-phenyl-thiorhodanide compounds (see Fig. 3.5.1). The NCS group at the terminal position strongly influences the properties. Such NCS containing compounds are well known for their decreased viscosity, which will lead to decreased switching times. NCS containing compounds are therefore very interesting for this work. In comparison with the alkyl chains at terminal position, the NCS group has in general the tendency to increase Δn at about 0.1. The used single compounds show a low melting temperature. The base matrix could be formed without strong restrictions, because the weight % of different compounds could be varied in a broad range.

The composition of the base NCS matrix (BMW1) is with (see. Fig. 3.5.1.) n=3,5,8, their weight percentages are 38%, 45%, 17% accordingly. The temperatures of phase transitions are

$$\text{Cr} < -17 \text{ N } 41,4 \text{ Iso}$$

The temperature of crystallization of this base matrix lies below -17°C and is predicted to be around -25°C. This matrix has a broad range of the nematic phase. The melting point below room temperature shows what is well known, namely one can prepare mixtures out of single compounds were each of them is with higher transition temperatures. The range of the nematic phase can be varied and, for example, if the component with n=8 have 10% and therefore the component with n=3 - 45%, the clearing point of the base matrix will be equal to 39,2°C. Otherwise increasing the weight ratio of the component with n=8 will lead to increased clearing point, but at the same time it will increase the temperature of melting. The value of $\Delta\varepsilon$ at room temperature is 7,65 at 1 kHz.

The values of the optical birefringence are influenced by the structure of the molecules. In this case there are two major factors. The presence of the cyclohexane ring instead of phenyl ring decreases the values of Δn, this fact is well known. On the other side as was stated above, the NCS group increases the birefringence. For the base matrix Δn is equal to 0.17, independent on the weight ratio of the components.

3.5.2.2 5CB Matrix

5CB (4-Cyano-4'-pentyl-biphenyl) is well known and its properties are widely investigated. This compound showed at room temperature a nematic phase, what is important for creating mixtures. 5CB was used as base matrix because this compound was already investigated in the microwave region and its anisotropy and dielectric losses were measured. By using 5CB as matrix for other additives we can extract the influence of these additives on the performance in the microwave region and therefore can improve mixtures. The chemical structure one can see under Add11.

The range of the nematic phase is moderate, phase transitions are

Cr 15°C N 36°C Iso

5CB has a CN group at the terminal position which is the reason for the high positive dielectric anisotropy. The measured value of $\Delta\varepsilon$ at room temperature is 12, the optical birefringence Δn is 0.18. The main contribution to the birefringence is due to biphenyl group, the CN group influence the birefringence only slightly. This compound has low rotational viscosity and is characterized by relative small switching times, what is important for the performance of the devices based on liquid crystals.

3.5.2.3 Tolane Matrix

Compounds used by us are out of the group of the presented Tolane molecules (see Fig. 3.5.1). In order to create a matrix we used three different compounds with n=3, m=2; n=3, m=4; n=4, m=4 (see Fig. 3.5.1.). The weight ratio of the compounds in this base matrix is 25%, 33%, 42% accordingly. The birefringence of the tolane 'base matrix' is with ca. 0,34 comparable high, these values are affected by the presence of a carbon-carbon triple bond [Goodby98].

The range of the nematic phase of the investigated compounds is even comparable high; therefore the created base matrix in the nematic phase is up to 140°C. The drawback of the tolane 'base matrix' is the limited UV stability, however it is not so important for experiments in the microwave region where samples and devices are sealed, as already stated.

3.5.3 Overview of the prepared mixtures based on the of NCS, 5CB and Tolane matrixes

On Table 3.5.1 the composition of all mixtures used in our experiments and their parameters at room temperature (phase transitions, birefringence, and dielectric anisotropy) are summarized.

Table 3.5.1 Parameters of BMW series of mixtures

Name	Composition (weight %)	Phase transitions	Δn	$\Delta \varepsilon$ (1kHz)
BMW1	NCS matrix 100%	Cr <-17°C N 41,4°C Iso	0,17	7,65
BMW2	NCS matrix 90% Add1 10%	Cr <-17°C N 46°C Iso	0,17	9,69
BMW3	NCS matrix 80%, Add1 12%, Add2 4%, Add3 4%	Cr 12°C N 54,9°C Iso	0,17	10,81
BMW4	NCS matrix 61%, Add4 13%, Add5 13%, Add6 13%	Cr <-17°C N 93°C Iso	0,22	7,51
BMW5	Tolane Matrix 85%, Add10 5%, Add9 10%	Cr <20°C N >130°C Iso	0,3	0,22
BMW7	Tolane Matrix 88%, Add1 5%, Add15 7%	Cr <20°C N >135°C Iso	0,32	8,42
BMW8	Tolane Matrix 85%, Add15 5%, Add16 10%	Cr -17°C N 46°C Iso	0,33	3,85
BMW10	Tolane Matrix 79%, Add17 7%, Add18 7%, Add11 7%	Cr 15C N >130°C Iso	0,33	6,25
BMW15	Tolane Matrix 80%, Add19 20%	Cr 13°C N >130°C Iso	0,32	4,56
BMW11	5CB Matrix 63,7%, Add7 31,3%, Add 8 10%	Cr <0°C N 89°C Iso	0,17	11,33
BMW12	5CB Matrix 85%, Add9 5%, Add10 10%	Cr -8°C N 40°C Iso	0,21	10,49
BMW13	5CB Matrix 80%, Add11 13%,	Cr -7°C N 50°C Iso	0,26	11,61

		Add12 7%			
	BMW14	5CB Matrix 80%, Add13 20%	Cr 5°C N 65°C Iso	0,26	12,85
	BMW16	5CB Matrix 55,5%, Add14 8,5%, Add7 26%	Cr <0°C N 99°C Iso	0,21	8,5

BMW2

Because of the presence of only 10% of Add1 the temperature behaviour is similar to BMW1. The weight ratio of the additive can be varied in a wide range, and our experiments show that even up to 25% of the weight ratio of this additive builds a eutectic mixture with the base matrix. The 10% ratio was chosen for preparation of BMW2 in order to keep the melting point of the mixture in the range of the melting point of the base matrix, thus below the 0°C. Because there is only one additive in this mixture, the comparison of the properties of both BMW1 and BMW2 can provide information about the properties of the single additive. In general, such method is widely used for the investigation of liquid crystals.

BMW3

For BMW3 the temperature of melting lies above room temperature. Our investigation shows that addition of Add3 to the base matrix or BMW2 does not build the nematic phase at room temperature, because this component crystallises by cooling down and no eutectic mixture is formed. Therefore Add2 was additional used, which has similar properties like Add1 but with a Fluoro group at the terminal position instead of a CN group. However even under this circumstance we were able to add only 4-5% of Add2. Increasing the weight ratio leads to the point where the mixture was unstable and the Add3 partially crystallizes at room temperature. Comparing the microwave performance with BMW2 we can see how such long molecules with an ethane fragment like Add3 influences the properties of the mixtures.

BMW4

BMW4 has very high N-Iso transition temperature in comparison with other mixtures based on the NCS matrix. This can be explained because the amount of the base matrix, which showed a clearing point at lower temperature, lasts only to 61%, so the clearing point of BMW4 is remarkable influenced by the additives. The weight ratio of the additives was chosen in order to keep the dielectric anisotropy of the mixture comparable to that of the base matrix BMW1 on one hand, but to improve the birefringence and to keep the melting point below 0°C on the other hand.

BMW5

Both additives Add9 and Add10 have Fluor atoms at lateral positions leading to a perpendicular dipole moment. The dielectric anisotropy of the base tolane matrix is positive, thus by mixing the components in proper ratio we can create a mixture with a dielectric anisotropy close to 0. So in the case of BMW5 we can check the microwave properties of the mixture with very low dielectric anisotropy and at the same time the influence of components with negative dielectric anisotropy.

BMW7

Because of the good properties of the matrix the weight percentage of both additives can be tuned in a broad range; the amount of additives was chosen so that the optical birefringence will not be changed very much. Experimental results confirms this, the value of birefringence for this mixture is equal to 0,32.

BMW8

The viscosity of the mixture is decreased compared to BMW7. The weight percentage of additives can be changed in a broad range and still components will form the nematic phase at room temperature. The exact composition of this mixture was chosen in order to obtain the birefringence like those of the matrix, but the dielectric anisotropy is still higher as those of the matrix.

BMW10

This matrix has components of different length; in particular one additive is a high temperature quaterphenyl. Because of Add11, it was important to include additives with low melting and clearing temperatures in order to create a mixture with the nematic phase at room temperature. For this purpose Add18 was used. In order to keep the birefringence close to the values of the matrix, addition of a higher amount of 5CB is not desirable and therefore only 7% of Add11 could be used.

BMW11

In comparison with the base matrix, BMW11 has very broad range of the nematic phase and the transition to the crystalline/glassy state is below room temperature. Such broad range of the nematic phase is due to Add7. The non polar Add8 is a low temperature nematics, like 5CB; together they build a very good eutectic mixture. Addition of 31,3% of Add7 shift the clearing point to higher temperatures. The amount of 31,3% was chosen in order to set the melting point of the mixture below RT. The viscosity of this mixture is higher compared to 5CB, because both additives contain a carboxylic group.

BMW12

The composition of this mixture was chosen in order to hold the nematic phase stable at RT, but improve Δn.

BMW13

Add11 is a high temperature component, because of this it is impossible to build binary eutectics with 5CB at RT, therefore a second additive Add13 was used for the mixture. The clearing point of this mixture can be increased by a higher amount of Add1. We also tried to add instead of Add11 the same quaterphenyl, but without C_5H_{11} alkyl chain and Fluor as lateral substituent. In this case it was not possible to create eutectics even with 5% of this additive.

BMW14

The amount of the additive in this mixture was chosen to keep this mixture in the nematic phase at RT. By increasing the amount of additive the mixture will crystallize at RT.

BMW16

The weight percentage of Add7 can be varied in a broad range. However this is valid for increasing of Add14 because of no nematic phase at RT.

The direct comparison with results for the pure 5CB can provide information about the influence of different molecules on the microwave performance. From mixtures BMW11, BMW14, BMW16 one can learn about the influence of the carboxylic group in view e.g. of viscosity and dielectric properties. Mixture BMW13 is characterized through the addition of components with very high optical birefringence in comparison with 5CB. Components in BMW12 have negative dielectric anisotropy, so one can study the function of laterally attached polar groups.

By comparing BMW1 with BMW2 and BMW3 the influence of increased dielectric anisotropy by keeping the birefringence constant can be studied. Moreover the boroxane ring and again the carboxylic group influences can be studied. Measurements of BMW3 can provide information about the influence of the extended 4-ring additive containing an ethane group. Also the increased viscosity of BMW3 is a point of interest in comparison with BMW2. Further, BMW4 has the same $\Delta\varepsilon$ like the base matrix but at the same time possesses enlarged birefringence. In this case we can expel the influence of the birefringence holding the dielectric anisotropy as constant.

3.5.4 Mixtures with high optical anisotropy.

The optical birefringence of the nematics should play an important role in the performance of the mixtures in the microwave region, therefore a set of mixtures with high birefringence were prepared. The best in this case would be the tolane matrix. The tolane matrix has a birefringence equal to 0,345. However that is not the highest possible value. As was introduced, quaterphenyl additives were chosen sometimes with compounds having a NCS group in the terminal position.

LHB13

This mixture consists of the tolane matrix and following components

with n=5; X=F for CompF1, n=5; X=OCF$_3$ for CompF2 and n=3 and X=F for CompF3.

The weight percentage of the tolane matrix and components is the following - 70,6 (Matrix) 5,7% (CompF1), 4,6% (CompF2), 2,1% (CompF3), 17% (Comp1).
This mixture has a nematic phase at room temperature with phase transitions
Cr 11° N >140° I.
The presence of components with F as lateral substituent allows us to use a high percentage of this additive in the mixture. The optical anisotropy of this mixture is lower than the values for CompF1-3, with birefringence equal 0,135. The anisotropy of Comp1 is 0,45. The measured

birefringence for LHB13 is equal to 0,325. The measured value of the dielectric anisotropy at the room temperature is 4,3.

LHB16

This mixture is based on the tolane matrix and the following components.

C_5H_{11} — [benzene]—[benzene with CH_3]—[benzene]—[benzene with F] — NCS   Comp2

C_3H_7 — [benzene]—[benzene with Cl]—[benzene]—[benzene with F] — NCS   Comp3

The weight percentage of the matrix and single compounds is 84%, 7%, 9% correspondingly. This mixture possesses nematic phase and is characterized via the phase transitions
Cr 8° N >140° I.

Only small amount of both additives can be added to the mixture in order to have nematic phase at room temperature. The NCS group (in both Comp2 and Comp3) in this mixture influence not only the birefringence but also increase the solubility of such long molecules in the matrix. Those should also decrease the viscosity of the mixture. The measured value of dielectric anisotropy is 3.5 at room temperature and 1kHz. Both Comp2 and Comp3 are characterized by high values of birefringence, the value of anisotropy is 0,44 and 0,45 correspondingly. Therefore the measured value of birefringence for the mixture is equal to 0,37.

LHB17

This mixture is based on the tolane matrix and following additives

C_5H_{11}—⟨benzene⟩—⟨benzene-CH$_3$⟩—⟨benzene⟩—⟨benzene-F⟩—NCS   Comp2

C_3H_7—⟨benzene⟩—⟨benzene-Cl⟩—⟨benzene⟩—⟨benzene-F⟩—NCS   Comp3

C_3H_7—⟨benzene⟩—⟨benzene-CH$_3$⟩—⟨benzene⟩—⟨benzene-F⟩—F   Comp4

The weight percentage of the matrix and Comp2, Comp3, Comp1 is 83%, 7%, 7%, 3% correspondingly.

This mixture has nematic phase and is characterized by phase transitions
Cr 10°C N >140° I.

The amount for Comp1 is chosen to obtain the nematic phase at room temperature, further the increase of the weight percentage of Comp4 will create the nematic phase only above room temperature. The measured value of optical anisotropy is equal to 0,365. The measured value of the dielectric anisotropy is equal to 3 at room temperature and 1kHz. The viscosity of the mixture should change only slightly in comparison to the pervious mixture.

LHB18

This mixture is based on the tolane matrix and following additives.

C_3H_7—⌬—⌬(CH$_3$)—⌬—⌬(F)—NCS   Comp1

C_5H_{11}—⌬—⌬(CH$_3$)—⌬—⌬(F)—NCS   Comp2

C_3H_7—⌬—⌬(Cl)—⌬—⌬(F)—NCS   Comp3

C_3H_{11}—⌬—⌬(Cl)—⌬—⌬(Cl)—NCS   Comp5

C_3H_7—⌬—⌬(CH$_3$)—⌬—⌬(F)—F   Comp4

The weight percentage of the matrix and Comp4, Comp2, Comp3, Comp5, Comp1 is 70%, 2%, 13,6%, 9,6%, 2,4%, 2,4% correspondingly. This mixture forms the nematic phase and is characterized by the phase transitions

Cr 18° N > 140° I.

The temperature of melting of this mixture is close to room temperature. This is because of the reasonable amount of the additives. The birefringence for this mixture is 0,395. The weight percentage of the additives was chosen in order to increase the birefringence. However these percentages are at maximum which can be achieved without raising the melting point above room temperature. Because all additives are quaterphenyls the viscosity of this mixture should be rather high.

3.5.5 Mixture with negative dielectric anisotropy

In order to check the influence of the direction and the strength of the dipole moment in the lateral direction, a mixture with negative dielectric anisotropy became prepared. The composition of such mixture BMW21 is presented below.

C_3H_7 — [three rings with CH$_3$, F, F substituents] — OCH_3 Comp6

C_3H_7 — [cyclohexane-phenyl-phenyl with F, F, F, F] — OC_2H_5 Comp7

C_nH_{2n+1} — [phenyl-phenyl-alkyne-phenyl with F, F] — C_mH_{2m+1} Comp8,9

The weight percentage of the components is 23,85%, 29,97%, 39,29% (Comp8: n=4, m=4) and 6,89% (Comp9: n=3, m=4) correspondingly. The phase transitions are:
Cr 38°C N 135°C Iso.

As was stated above the molecules with Fluor atoms at the lateral positions have negative dielectric anisotropy (Comp6,7). At the same time tolane compounds (Comp8,9) in this mixture are weakly polar components. Therefore the measured dielectric anisotropy for BMW21 at 1 kHz and room temperature is -1.5.

4.0 Results and discussion

4.1 Ferroelectric LC's – X-Ray and microwave measurements.
4.1.1 X-Ray diffraction (XRD) of FLC.

As was shown in the Experimental part [Chapter 3.2.2] for microwave investigations FLC mixtures with a high tilt angle, preferable close to 45°, are needed in order to direct the highest component of the optical anisotropy as a working parameter. Here results of XRD measurements on selected FLC compounds will be presented. Using XRD we can determine the averaged tilt angle of the whole molecules. In contrary electro-optical measurements detect a tilt angle which is to a great extent related to the core of the molecules only. In general, in the smectic C/smectic C* phase, the core is more tilted than the whole molecule [Watson02]. If one can find the ratio between the optical tilt angle and the tilt angle based on XRD measurements for a certain class of compounds then one can estimate the 'optical' tilt angle out of X-Ray data.

For several compounds (which will be presented later on) showing the chiral smectic C* phase, selected physical properties (layer thickness and 'XRD' tilt angle, temperature range of the SmC* phase, spontaneous polarization) were investigated (Table 4.1).

$H_{21}C_{10}$–⟨⟩–⟨⟩–COO–⟨⟩–⟨⟩–COOCH(CH_3)C_6H_{13}
  |
  Cl
                                                                          SC-5

$H_{21}C_{10}$–⟨⟩–⟨⟩–COO–⟨⟩–⟨⟩–COOCH(CH_3)C_6H_{13}
  |
  H_3C
                                                                          SC-48

$H_{21}C_{10}$–⟨⟩–⟨⟩–⟨⟩–OOC–⟨⟩–OCH(CH_3)C_6H_{13} (S)
  |
  H_3C
                                                                          SC-90

         CH_3
         |
$H_{21}C_{10}$–⟨⟩–⟨⟩–⟨⟩–OOC–⟨⟩–OCH(CH_3)C_6H_{13} (S)
                                                                          SC-91

$H_{21}C_{10}$–⟨⟩–⟨⟩–⟨⟩–OOC–⟨⟩–OCH(CH_3)C_6H_{13} (R)
  |
  H_3C
                                                                          SC-94

H₁₇C₈―⟨⟩―⟨⟩―⟨⟩―OOC―⟨⟩―OCH(CH₃)C₆H₁₃ (R)
   H₃C

SC-97

H₂₁C₁₀―⟨⟩―⟨⟩―⟨⟩―OOC―⟨⟩―OCH(CH₃)C₆H₁₃ (R)
   Cl

SC-98

The XRD tilt angle can be estimated from measuring the layer thickness (d_{SmC}) in the SmC phase and by calculating the length (l_{mol}) of molecule. $Sin\Theta = l_{mol} / d_{SmC}$.

If the sample has both SmA and SmC phases it is possible to calculate the crystallographic tilt angle also by using the layer thickness in SmA phase instead of molecular length.

TABLE 4.1 Electrooptical and X-ray properties of single compounds (SC)

Mixtures	SmC* temperature range [°C]	Spontaneous polarization [nC/cm²]	'Optical' Tilt angle [°]	d [Å] in SmC*	Tilt angle extracted from X-ray data [°]
SC-5	37 - 142	160	44	31.2	39.6
SC-48	40 - 115	140	36.8	33.6	31.4
SC-90	< 20 - 113.5	175	35.7	28.4	28.1
SC-91	(70.2)	-	37.1	26.5	29.6
SC-94	< 20 - 123.6	170	35.8	28.4	28.1
SC-97	<20 - 110.3	172	37	25.6	29.0
SC-98	<20 - 132.4	175	38.8	27.5	31.0

Compounds SC-5 and SC-48 are different only on the first phenyl ring close to the alkyl chain by substituting Cl via CH₃ group resulting in a remarkable high tilt angle in favor of the lateral Chloro group. The same is true by considering the pair SC-94 and SC-98.

All those four single molecules are with a long chiral group on one side, a long alkyl chain on the other side and a comparable long core built from four phenyl groups with a carboxylato group in between. In general, compounds with a carboxylato group in the core showing improved tilt angles. So by combining the carboxylato group in the core with a lateral Chloro group the 'Optical' tilt angle reaches close to 45°. The great difference up to 8° between

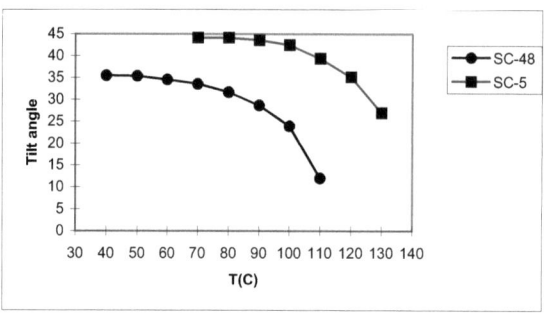

Figure 4.1.1 Temperature dependence of the tilt angle for SC-5 and SC-48

'Optical' tilt angle and 'X-ray' tilt angle is evident. The temperature dependence of the tilt angle for SC-5 and SC-48 is shown in Fig. 4.1.1. [V.Lapanik04].

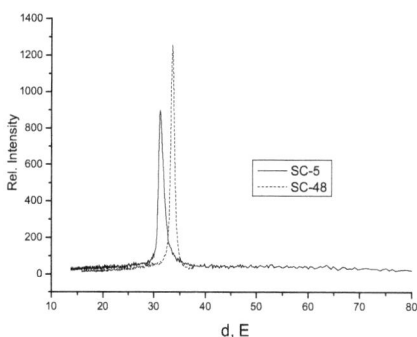

Figure 4.1.2 XRD pattern for SC-5 and SC-48 at 75°C

The XRD measurements of these FLC compounds confirm the electro-optical results. (Tab. 4.1). The thickness of the SmC* layers for SC-5 is smaller than for SC-48. Since both compounds have the same molecular length, the tilt angle in case of SC-5 is enlarged, compared to SC-48 (see Table. 4.1.1) what again says, introducing the lateral Cl atom leads to the increase of the tilt angle.

In continuation of the presented investigation we have investigated one other pair of FLC compounds (SC-91 and SC-90). The difference between them is only the position of the CH_3 substituent in the central core. Both single compounds are left-handed with the same molecular length. X-Ray measurements have shown that the thickness of the SmC* layer for

SC-91 is lower than for SC-90 (see Table 4.1 and Fig. 4.1.3). It may be explained by a different packing of the molecules inside the layer, leading to the different tilt. It should be noted that these results are in agreement with results from electro-optical measurements (see Table 4.1).

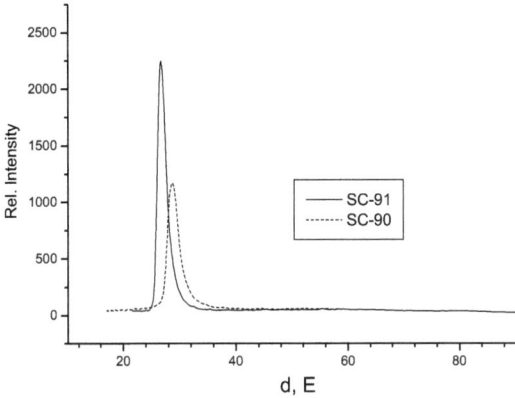

Figure 4.1.3 XRD pattern for SC 90 and SC 91 at 57°C

The influence of the length of alkyl chain on the tilt angle can be demonstrated by comparing the XRD results for SC-94 with a decyl group and SC-97 with an octyl group. The results of the XRD and the electro-optical measurements for SC-94 compared to SC-97 demonstrated that the tilt angle of the compound with a shorter alkyl chain (SC-97) is higher in comparison to SC-94 (see table 4.1). Because similar results we received for other comparable single compounds, one can generalize: Decreasing the length of the alkyl chain results in higher tilt angle.

For single compounds and to some extent for mixtures one can predict by knowing the structural formula the tilt angle regardless if XRD data or optical data are considered. This information has some consequence following the aim of the work to design and optimize mixtures as good candidates for microwave investigations.

4.1.2 Microwave measurements of FLCs

In comparison with nematic mixtures, FLC mixtures are characterized by low switching times (1μs-500μs). Therefore FLC mixtures are very interesting for microwave applications, where the switching time for some devices is an important parameter. In scope of this work, different FLC mixtures were prepared, but only one mixture could be characterized in the microwave region.

The main problem with respect to the use of FLCs for the microwave purpose is the quality of orientation at thicknesses close to 25 μm. It is well known that typical FLC mixtures can not be oriented with good enough quality at such thicknesses. Therefore some mixtures were prepared with the aim to overcome this problem. One of them is the mixture LAHS7 with the composition:

Three phenylpyrimidin compounds with 80% weight percentage (n=8, m=6 – 35%; n=8, m=8 – 30%; n=10, m=8 – 35%)

$H_{2n+1}C_n$ —[pyrimidine]—[phenyl]— OC_mH_{2m+1}

-Two chiral compounds – both components have 10% of weight percentage

$C_4H_9(CH_3)C^*HOC(O)$ —[phenyl]—[phenyl]—[phenyl]— $C(O)OC^*H(CH_3)C_4H_9$

$H_{13}C_6C^*H(F)CH_2O$ —[phenyl]—[pyrimidine]— $OCH_2C^*H(F)C_6H_{13}$

The temperatures of the phase transitions are (by heating up)
Cr 5°C SmC* 47°C SmA* 56°C N 73°C Iso.
This mixture is characterized by rather low value of spontaneous polarization ~10nC/cm^2. The tilt angle of this mixture is equal to 25°; one can note that this value is far below the maximum of 45° for smectics. That means that the anisotropy of the mixture can't be used optimally, mean the effective $\Delta\varepsilon$ of the mixtures with tilt angle lower than 45° is lesser [Chapter 3.2.2]. On figure 4.1.4 the results of broadband measurements of LAHS7 and 5CB are presented [Goelden07].

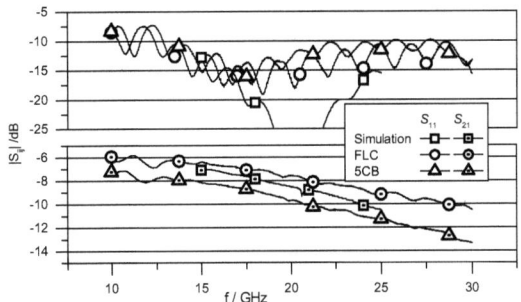

Figure 4.1.4 S-parameters for 5CB and LAHS7

Scattering parameters or S-parameters are properties used in electrical engineering, describing the electrical behaviour of linear electrical networks when undergoing various states. S-parameters are a mathematical subset of a more general non-linear formulation called X-parameters [Pozar05]. S11, input return loss, is equivalent to the reflected voltage magnitude divided by the incident voltage magnitude. S21 is the gain; the voltage gain is a linear ratio of the output voltage divided by the input voltage. Insertion losses (S11) are lower in the case of LAHS7 in comparison with 5CB. The material parameters could not be extracted from these results; however the difference in S11 parameter shows that FLC can have lower losses than typical nematic mixture 5CB. On Fig. 4.1.5 the FOM for these two mixtures is presented.

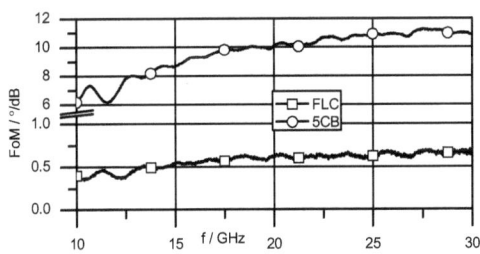

Figure 4.1.5 Figure of Merit of 5CB and LAHS7

One can see that 5CB shows far better performance in comparison with LAHS7. The reason should be the difference in the effective dielectric anisotropy at these frequencies. As was stated above the value of the tilt angle is lower than optimal and at the same time the geometry of the testing cells doesn't use the anisotropy of LC molecules fully in comparison with measurements that could be made with coplanar waveguides and mixture with tilt angle close to 45° [Moritake]. As a result there is a need to improve the performance of the devices, in particular for FLC mixtures. Only in terms of dielectric losses at these frequencies LAHS7 looks good in comparison with 5CB.

LAHS7 was also investigated at several frequencies in order to check how the performance of the mixture changes depending on the frequency. On Fig. 4.1.6 the results of these measurements are presented.

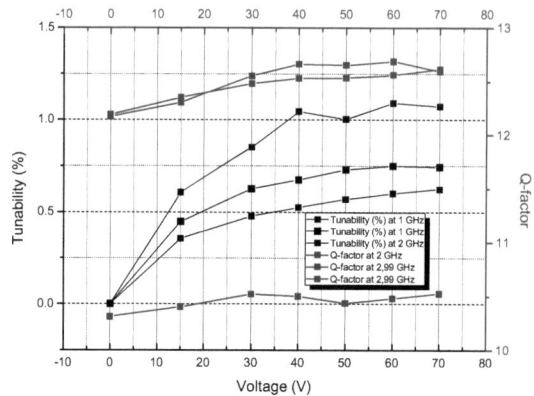

Figure 4.1.6 Performance of LAHS7 at GHz frequencies

The Q-factor is defined as $\Phi/\tan \delta_{\varepsilon, max}$, where Φ is the phase shift and $\tan \delta_{\varepsilon, max}$ is the maximal value of losses. There is the same tendency as for nematics - with the increase of frequency the losses become lower [Weil03]. The quality factor at 3 GHz has the maximum value. Higher losses at this region might be explained via the high frequency molecular mode at the 1-100 MHz region, which can still greatly affect the 1-10GHz region [Utsumi04]. But this is still under discussion.

4.2 Nematic mixtures at MHz region.

4.2.1 Mixtures based on the NCS matrix

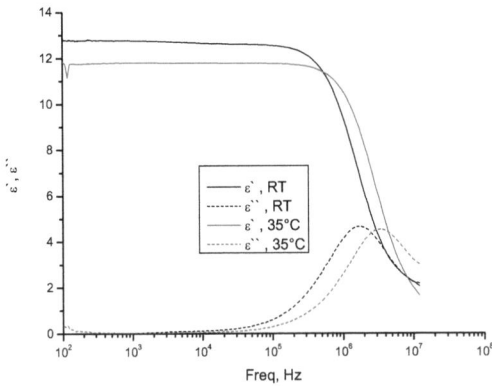

Fig. 4.2.1 Dielectric spectra of BMW1

The dielectric spectrum of BMW 1 is presented in Fig 4.2.1. At room temperature up to 2 absorption processes are feasible. The small absorption in the high frequency range around 10 MHz is caused by the absorption (RC constant) of the measuring cell. The cell used for BMW 1 showed the same RC parameter as all cells used in this series; mean the absorption at about 10 MHz is always visible. The relaxation process in the range of 1MHz can be identified as molecular mode and more precisely as reorientation around the short axis of molecules. This process took place because the components of the mixture have a longitudinal dipole moment and the orientation of the mixture in the cell is homeotropic. The frequency of the process increases with temperature what is a normal behaviour of this activated molecular mode. The absorption intensity is rather high (e"~ 5) because of the strong dipole moment formed by the NCS group.

Figure 4.2.2 shows the dielectric spectra at different temperatures for BMW2.

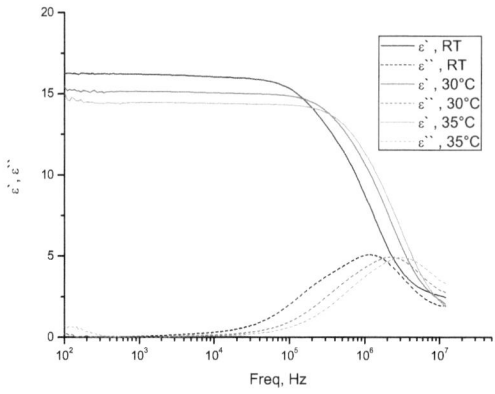

Fig. 4.2.2 Dielectric spectra of BMW2

This spectrum is similar to that of BMW1 what is reasonable because the same base matrix with a high weight ratio is used. The temperature activated molecular mode, interpreted as above as diffuse reorientation around the molecule short axis, is slightly above 1MHz. At room temperature a weak process appears as shoulder at 100 kHz on the lower frequency side of the strong absorption. With increase of temperature this absorption disappears or became overlapped with the higher frequency molecular mode. This process can be identified as relaxation of one of the components in this mixture, probably the largest/heaviest one. To assign this process one should look on the components of the mixture. Here, the molecular shape of the boroxane molecule Add1 is to some extent different from the shape of the remaining components of the mixture. Probably this molecule will fulfil its relaxation around the short molecular axis at lower frequency compared to the relaxation of the major components of the mixture (BMW1). The intensity of the lower frequency relaxation is because of only 10 % in weight less compared to the higher frequency process. The discussed two processes can be clearly seen on the Cole-Cole plot in fig. 4.2.3

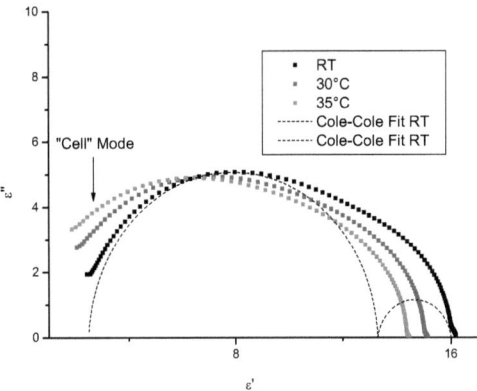

Fig. 4.2.3 Cole-Cole plot of BMW2

The alpha parameter of the low-frequency process obtained by fitting of the Cole-Cole equation is 0,1 while the critical frequency is 250 kHz at room temperature. The high frequency process showed an alpha parameter of 0,03. This relaxation is a quasi Debye type relaxation. The frequency of this process is 1,176MHz at room temperature.

The dielectric spectrum of BMW3 is presented in figure 4.2.4.

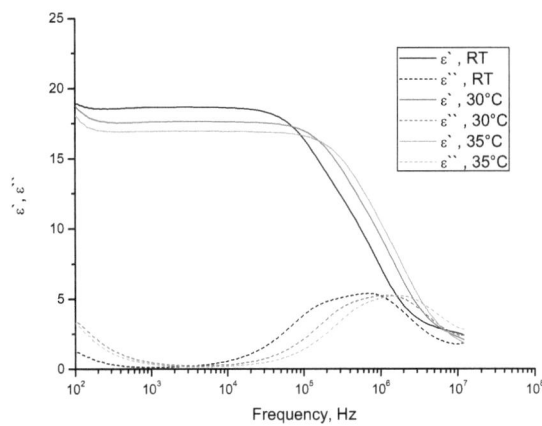

Fig. 4.2.4 Dielectric spectra of BMW3

The molecular mode (rotation around the short molecular axis) for BMW3 is characterized by lower critical frequencies (at room temperature below 1MHz) compared to the former mixtures. The reason for this is the increased rotational moment around the short axis of the entire system due to the use of 3 ring components Add1 and Add2 with a COO bridge fragment and one 4 ring component with an ethane bridge Add3. As before there is an indication for another mode that lies close to the major molecular mode. Even this mode is more prominent at room temperature, and with increased temperature the critical frequency shifts to the higher frequency range. If we compare the spectrum of BMW3 with the dielectric spectrum of BMW2 we can see that this lower frequency rotational mode behave approximately by the same frequency as in BMW2. Because both mixtures contain the same boroxane molecule Add1, one can assign as before this mode as the characteristic relaxation of this boroxane component. In comparison to the mixture BMW2, the mixture BMW3 contain a lower weight ratio of the base matrix and thus a comparable higher amount of the larger molecule leading to a more prominent additional relaxation mode compared to BMW2.

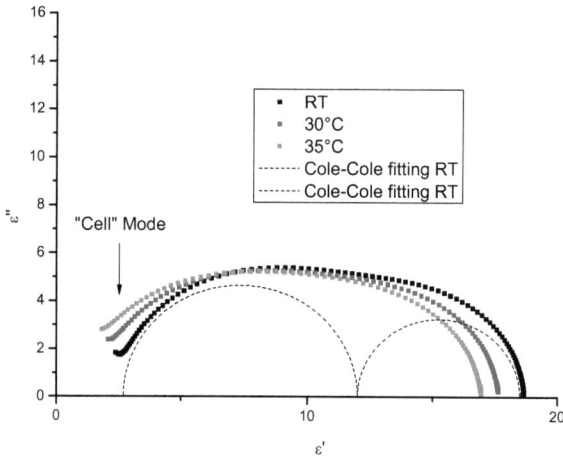

Fig. 4.2.5 Cole-Cole plot of BMW3

The low frequency mode was fitted to the Cole-Cole equation, the alpha parameter was found to be 0. That means a Debye type relaxation without any spreading of the spectrum take place. The frequency of this process is 160 kHz at room temperature. The high frequency mode is also a Debye type relaxation at 670 kHz. One should note that the low dispersion mode in this case is also Debye like, as for BMW2.

The dielectric spectrum of BMW4 is shown in figure 4.2.6

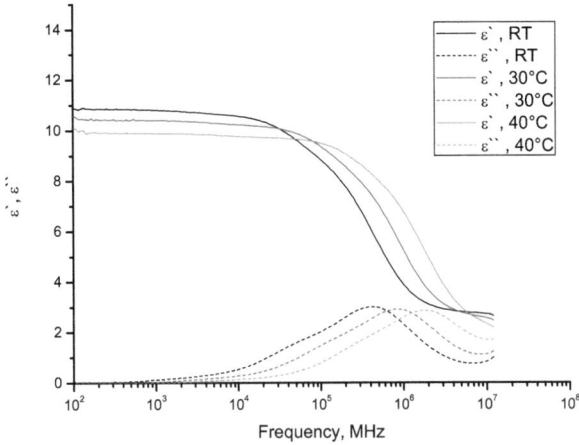

Fig. 4.2.6 Dielectric spectra of BMW4

The molecular mode which corresponds to the rotation of the molecules around the short axis has lower critical frequencies in comparison to the other mixtures of this series because of the larger rotational moment of the molecules in this mixture with ethane bridge fragments Add 4 and Add5. The decreased dielectric anisotropy, caused by the component with a perpendicular dipole moment Add6, lead to a decreased intensity of the molecular mode and thus to decreased losses. Also this mixture shows an absorption process on the lower frequency side of the main molecular mode caused by components with ethane bridge fragments. At higher temperatures this process is more overlapped by the main molecular mode in the higher frequency range.

Fig 4.2.7 presents the Cole-Cole plot. The lower frequent relaxation mode has a critical frequency of 45 kHz, which is a very low value. The only possible conclusion is, this process belongs to additives with ethane fragments (Add4,5) because of the high moment of inertia.

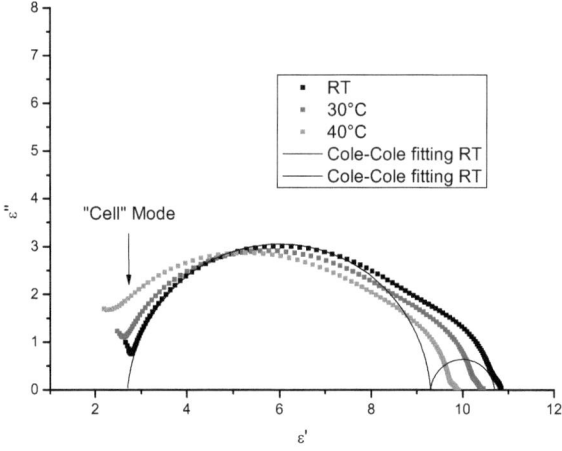

Fig. 4.2.7 Cole-Cole plot of BMW4

The alpha parameter of the Cole-Cole equation is 0,06. The high frequency mode appears at a critical frequency of 410 kHz with an alpha parameter (Cole-Cole) of 0,05.

Figure 4.2.8 shows the critical frequencies of the main absorption processes for the mixtures BMW1 to BMW4 in dependence of the temperature.

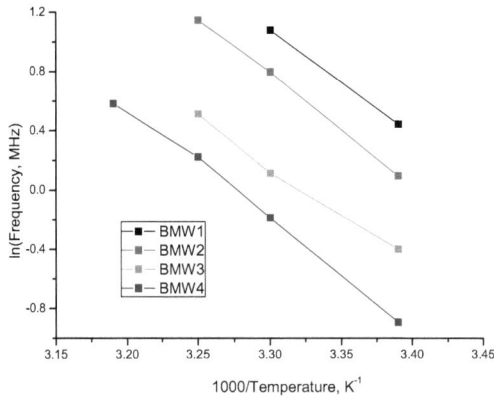

Fig. 4.2.8 Arrhenius plot for mixtures BMW1 to BMW4 based on the NCS matrix

The activation energies for these processes were calculated. For BMW1-4 Ea is equal to 31kJ/mol, 33kJ/mol, 29kJ/mol, 32kJ/mol correspondingly. This is a rather expectable result. The activation energies have almost the same values for BMW1-4 mixtures because the NCS matrix has very dominant weight percentage in these mixtures.

4.2.2 Mixtures based on the 5CB matrix

Fig 4.2.9 presents the dielectric spectra for the BMW11 mixture.

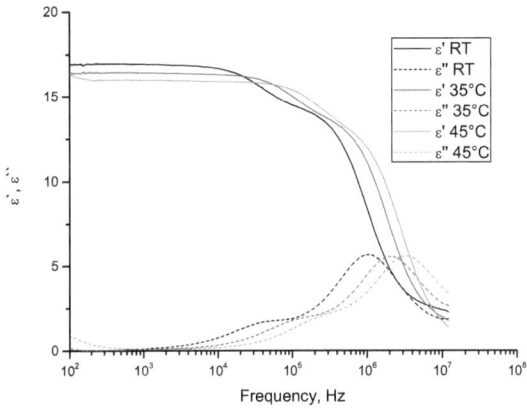

Fig. 4.2.9 Dielectric spectra of BMW11

Three areas of absorption processes can be seen. The high frequency process is the tail of the RC resonance. At 1 MHz the diffusive reorientation around the short axis of molecules take place. This molecular mode shows clearly the dependence on temperature indicating the positive activation energy. The dielectric properties of 5CB are well known and from several research groups investigated. One can find slightly different values for the reorientation of 5CB around the short axis for example 25MHz [Belyev99], 15,6MHz [Gu07], in any way above 10MHz. In the case of BMW11 and other mixtures in the 5CB series this relaxation mode is shifted to a frequency range which takes place of one order lower. This is due to the influence of the other compounds in the mixture. The process around 100 kHz has lower intensity in comparison with the process at 1 MHz. This process is related to the molecular mode of the Add7 in this mixture, because this component showed a higher value of the rotational moment. The two molecular modes can be separated in a Cole-Cole plot, shown on fig. 4.2.10

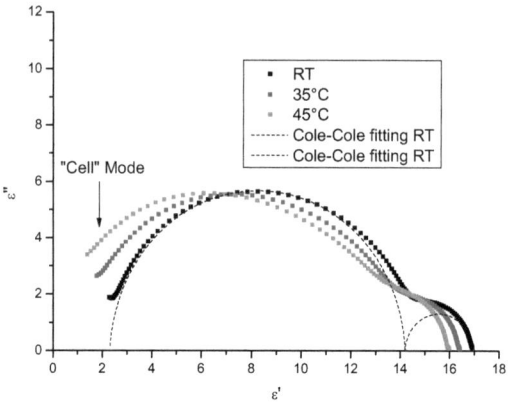

Fig. 4.2.10 Cole-Cole plot of BMW11

For the low frequency mode the alpha parameter is equal to 0,03 which is very close to the Debye type relaxation. The critical frequency of this process is at 46 kHz. For the high frequency process the alpha parameter is equal to 0,02. The critical frequency is 1,02 MHz.

The dielectric spectrum of BMW 12 is shown in fig. 4.2.11. This mixture contains only two additives with high negative values of dielectric anisotropy.

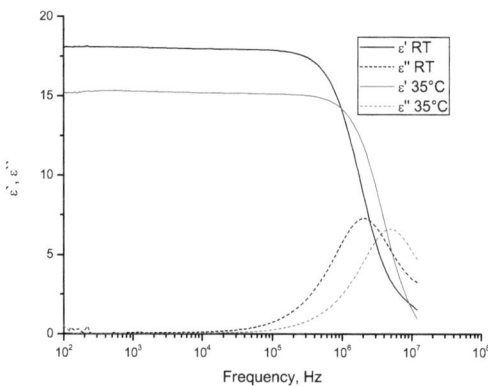

Fig. 4.2.11 Dielectric spectra of BMW12

Because the reorientations around the long molecular axis as a consequence of perpendicular dipoles take place at much higher frequencies (close to 1 GHz), the presented spectrum of

BMW12 in Fig.4.2.11 is more comparable with them of the pure 5CB in a matrix. The critical frequency of the molecular process in case of BMW12 is slightly higher as for BMW11 because of the obviously decreased rotational moment of molecules. There is no indication of a molecular mode of the single additives in the mixture in the investigated frequency range what can be easy explained by considering their negative dielectric anisotropy. The critical frequency of this process is equal to 2,08 MHz. The Cole-Cole fitting results in 0,1 as alpha parameter.

Fig 4.2.12 shows the dielectric spectrum for BMW13.

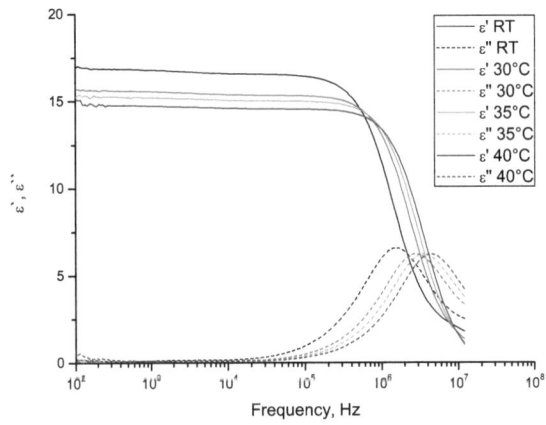

Fig. 4.2.12 Dielectric spectra of BMW13

The absorption strength of the process at 1 MHz is higher than for previous mixtures on the base of 5CB because here both additives (Add11,12) possess the dipole moment in the longitudinal direction. There is no indication of a second molecular process. This is interesting because the 4 ring additive (Add11) show a strong extension in the length in comparison with 5CB.

From the Cole-Cole plot one can see too there is only one relaxation process (see Fig. 4.2.13)

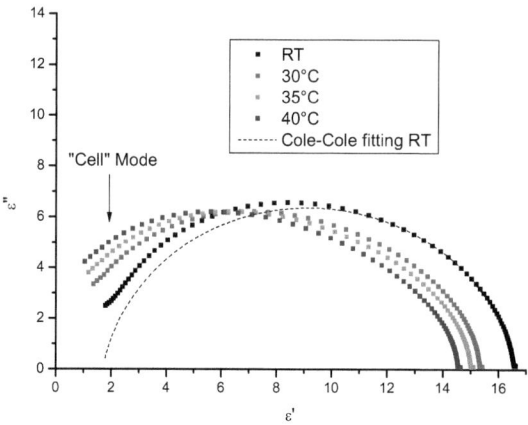

Fig. 4.2.13 Cole-Cole plot of BMW13

This single process has a critical frequency at 1,56 MHz. The alpha parameter is 0,1.

In Fig 4.2.14 the dielectric spectrum of BMW14 is presented. Because the mixture consists only of two components and they are considerable different in shape and length, one could expect that both components will have their own critical frequencies of the molecular relaxation mode.

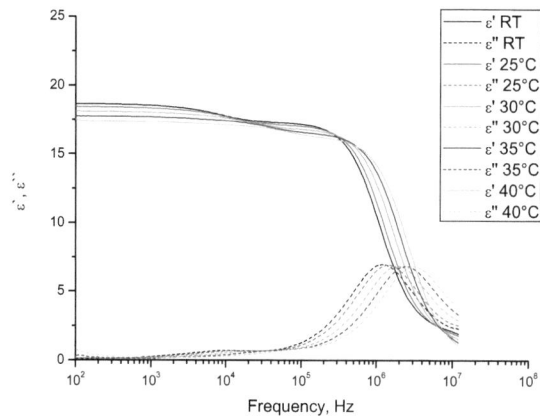

Fig. 4.2.14 Dielectric spectra of BMW14

Indeed, this mixture shows two relaxation modes which can be identified as rotation around the short axis of different molecules. As in the former cases the process around 1 MHz belongs to 5CB molecules, the lower frequency process to the Add13 component. The Cole-Cole plot is presented in Fig. 4.2.15.

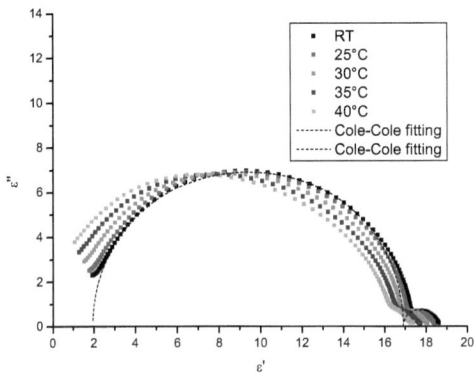

Fig. 4.2.15 Cole-Cole plot of BMW14

The intensity of the molecular mode of the additive is much weaker compared to them of the base component 5CD, this is in full agreement with the low weight ratio of the Add13 in the mixture. The critical frequency of this process is only at 12 kHz, the alpha parameter 0,1. The high frequency mode should correspond to 5CB. Its critical frequency is 1,18 MHz, close to the values of this mode in previous mixtures on the base of 5CB. The alpha parameter is equal to 0,05.

The dielectric spectrum for mixture BMW16 is presented in Fig. 4.2.16.

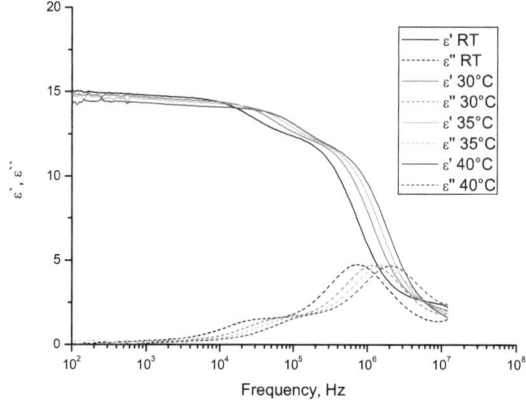

Fig. 4.2.16 Dielectric spectra of BMW16

This mixture again consists of molecules with large differences in shape and length. Therefore there are two molecular modes on this spectrum, one mode at around 1MHz (5CB), the second between 10 kHz and 100 kHz. The last process equals to the rotation around the short axis of the Add14 with broken rigidity because of the COO bridge fragment. The big value of the dielectric anisotropy for this molecule causes a relative strong intensity of the lower frequency molecular mode even if the weight ratio of this molecule in the mixture is low. The high frequency range is also affected by absorption processes caused by the RC resonance.

The Fig 4.2.17 presents Cole-Cole plot, which shows clear the separation between two molecular modes.

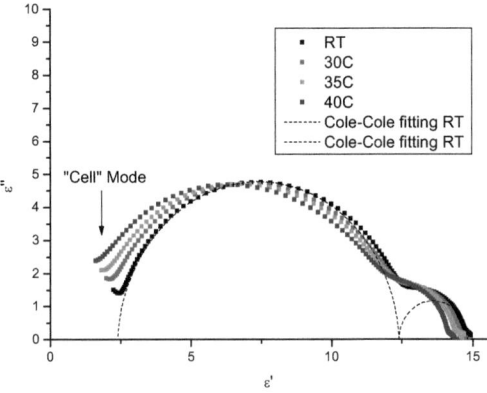

Fig. 4.2.17 Cole-Cole plot of BMW16

The critical frequencies of these modes are 30 kHz and 714 kHz. Both modes are described by Cole-Cole equation with α=0,03.

4.2.3 Mixtures based on the Tolane matrix

Fig 4.2.18 presents the dielectric spectra of BMW 5.

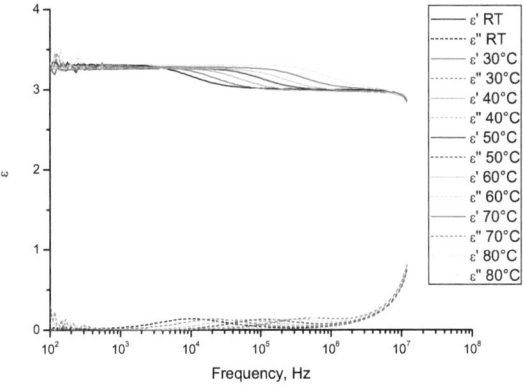

Fig. 4.2.18 Dielectric spectra of BMW5

There is only one very weak process in the range of 10 kHz at room temperature, the frequency increases with temperature. This proves that BMW5 has a dielectric anisotropy close to 0 and therefore the molecular mode, the rotation around the short axis, shows small intensity. Due to the fact that molecules in the tolane matrix and additives with negative dielectric anisotropy have similar length, there is no indication of separating the molecular process into individual contributions of different single molecules. This can be seen on Fig 4.2.19, which presents the Cole-Cole plot. ε" is in the range of only 0,1-0,2. Remarkable, even the high frequency process which corresponds to the mode of the gold cell has much higher intensity.

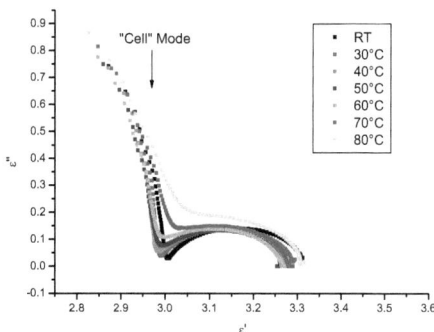

Fig. 4.2.19 Cole-Cole plot of BMW5

On fig. 4.2.20 the dielectric spectra for the BMW8 mixture is presented.

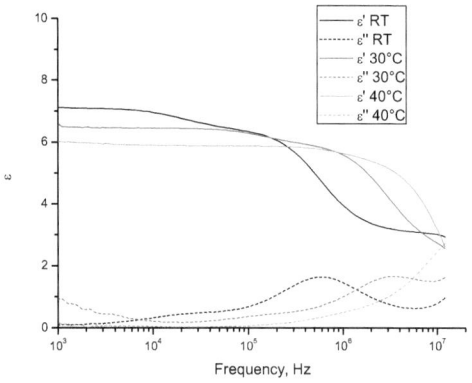

Fig. 4.2.20 Dielectric spectra of BMW8

Again, there is only one very weak process in the range of 1 MHz at room temperature, the critical frequency increases with temperature. In comparison with BMW5 this mixture has higher values of positive dielectric anisotropy; it contains the small length boroxane component (Add15). From Cole-Cole plot at fig 4.2.21 one can see another molecular mode at lower frequencies, but the intensity of this mode is very small.

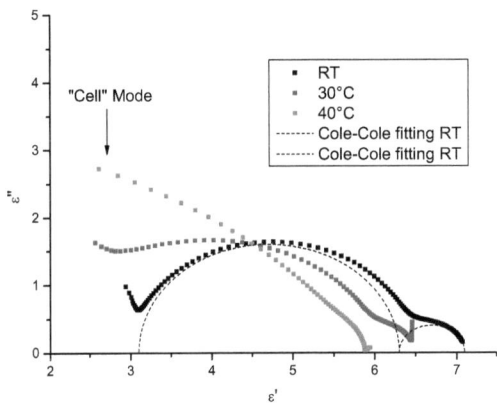

Fig. 4.2.21 Cole-Cole plot of BMW8

The presence of this mode (critical frequency is 40 kHz, $\alpha=0$) is rather unclear because there is no strong deviation of the shape of the molecules in this mixture. The higher frequency mode appears at 580 kHz, also described by a Debye type function.

The dielectric spectrum of the BMW10 mixture is to some extent different in comparison to the other mixtures with long additives. (see Fig. 4.2.22)

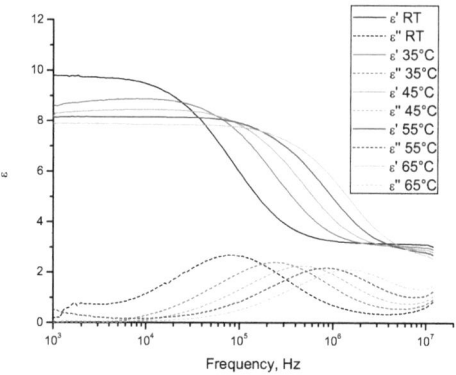

Fig. 4.2.22 Dielectric spectra of BMW10

There is again a strong temperature dependence of the molecular process, for instance at RT the critical frequency is at ~100 kHz whereas at 45 °C this belongs to ~1 MHz. One could except that because of differences in the length of the matrix and single additives several molecular processes would appear. However the Cole-Cole plot indicates only one absorption process (see Fig. 4.2.23).

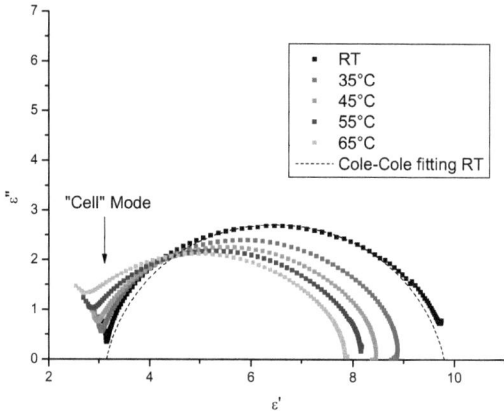

Fig. 4.2.23 Cole-Cole plot of BMW10

The presence of quaterphenyl additive (Add11) shifts the frequencies of the molecular mode to lower values. The critical frequency of this mode is 81 kHz and $\alpha=0,13$.

4.3 Nematic mixtures at 30GHz

4.3.1 Mixtures based on the NCS matrix

All mixtures of the BWM series based on the lateral group NCS are characterized in the microwave region by the cavity perturbation method [Goelden07a]. Measurements were done at 30 GHz depending on temperature. The results of these investigations are presented below. Components of dielectric permittivity and losses are shown.

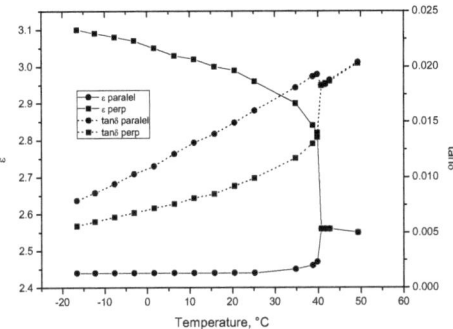

Fig. 4.3.1 Losses and permittivity versus temperature for BMW1

On fig 4.3.1 we can see the values of permittivity and losses, both perpendicular and parallel (to the long axis of molecules), versus temperature for the BMW1 mixture. Both values show strong temperature dependence, except the perpendicular component of the permittivity. The best values of losses and dielectric anisotropy at this frequency behave at lower temperature which is due to the higher order parameter at lower temperatures. The clearing point of this mixture is at 41,4°C. Around this temperature both permittivity components become equal. The dielectric losses at room temperature are $\tan\delta_\parallel = 0,009$ and $\tan\delta_\perp = 0,015$. It should be noted that BMW1 has a broader range of the nematic phase and is not crystallizing down to -17°C.

The fig 4.3.2 shows components of dielectric permittivity and losses for the mixture BMW2. This mixture behaves as BMW1. At room temperature the dielectric anisotropy of both mixtures are comparable. Moreover, both mixtures show the same optical birefringence. However the losses are quite more prominent for BMW2. The parallel component is slightly different, but this is in the range of experimental error. However $\tan\delta_\perp$ is higher. Our measurement shows that both components of permittivity become equal at temperature between 40°C and 50°C indicating the clearing point at 46°C. Like BMW1 this mixture did not crystallize down to -17°C. The dielectric losses at room temperature are $\tan\delta_\parallel = 0{,}01$ and $\tan\delta_\perp = 0{,}018$.

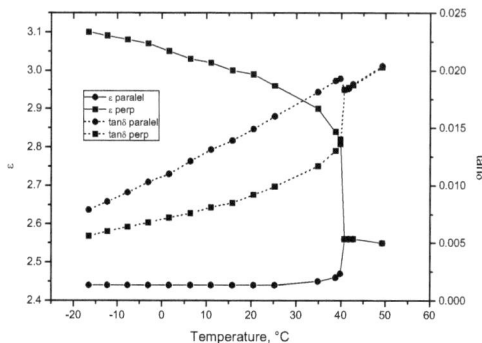

Fig. 4.3.2 Losses and permittivity versus temperature for BMW2

Fig 4.3.3 shows the results of measurement for mixture BMW3. The behaviour of this mixture is different from previous ones because the dielectric anisotropy and the losses change close to 10° C. This can be explained as partial crystallization at this temperature. Close to the temperature of -17°C the anisotropy suddenly increases and at the same time losses become also higher. This behaviour is most probably a consequence of the partial oriented glassy phase.

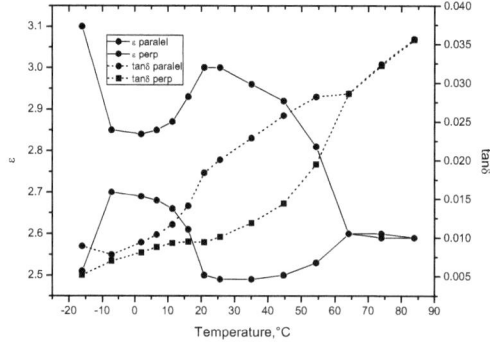

Fig. 4.3.3 Losses and permittivity versus temperature for BMW3

The dielectric losses at room temperature are $\tan\delta_\parallel = 0{,}009$ and $\tan\delta_\perp = 0{,}018$. In this sense this mixture shows the same performance at room temperature as BMW2. The value of dielectric anisotropy stays at the same level, because this mixture has the same Δn as previous mixtures. Both components of permittivity become equal at the temperature between 55°C and 65°C indicating the clearing point at 54,9°C.

With fig 4.3.4 data for BMW4 are presented. BMW4 has the same temperature dependence of parameters like all mixtures based on the NCS matrix. However because of the higher value of the optical birefringence this mixture also has higher dielectric anisotropy. The range of the nematic phase is rather broad; no crystallizing down to -17°C appears. The clearing point is at 93°C, at this temperature (92°C and 97°C) both components of permittivity become equal. The dielectric losses at room temperature are $\tan\delta_\parallel = 0{,}008$ and $\tan\delta_\perp = 0{,}018$. Remarkable, the parallel component of losses for this mixture is higher than for the previous NCS mixture; however this increase is very small and can be neglected.

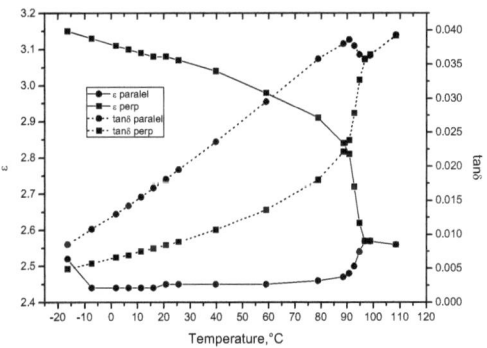

Fig. 4.3.4 Losses and permittivity versus temperature for BMW4

All mixtures of the BMW series (with longitudinal NCS compound) show almost the same performance at 30 GHz. The value of the optical birefringence Δn for the first three mixtures is comparable and as a consequence the dielectric anisotropy in the microwave region is comparable too. The forth mixture has higher Δn and as result $\Delta\varepsilon$ is also higher. The values of the parallel component of the optical permittivity of all mixtures are almost the same which is in good agreement with the dominant longitudinal dipole moment of the NCS containing molecules in a high weight ratio in all mixtures.

In the following table the values of τ and FOM at room temperature for all NCS mixtures are presented.

Table 4.3.1 Microwave properties of BMW1-4 mixtures

	BMW1	BMW2	BMW3	BMW4
Tau	0.18	0.18	0.17	0.21
FOM	12.24	9.94	9.23	11.40

One can see that because of lower losses for the perpendicular component, BMW1 show the highest FOM and therefore the best microwave performance of this series of mixtures. Because of the higher dielectric anisotropy BMW4 has better values of FOM than BMW2 and BMW3.

On Fig 4.3.5 the Figure of Merit (FOM) versus temperature for all mixtures is presented.

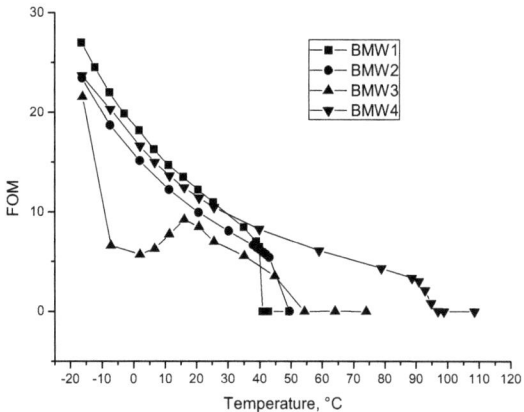

Fig. 4.3.5 Figure of Merit for BMW1 to BMW4 NCS mixtures

4.3.2 Mixtures based on the 5CB matrix

All mixtures on the base of 5CB are characterized with help of the microwave cavity perturbation method. Temperature dependent measurements were performed at 30 GHz. The results of these investigations for some of the mixtures are presented below. Components of dielectric permittivity and losses are shown.

On the Fig 4.3.6 we can see the value of permittivity and losses, both perpendicular and parallel, versus the temperature for the BMW11 mixture.

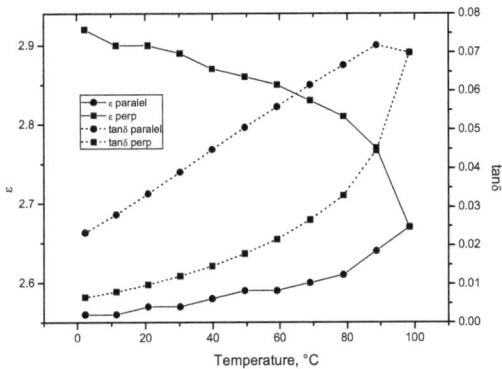

Fig. 4.3.6 Losses and permittivity versus temperature for BMW11

Both parameters have strong temperature dependence, except the perpendicular component of the permittivity. The best values of losses and dielectric anisotropy at this frequency are at lower temperatures, which is a consequence of the increased order parameter. The clearing point of this mixture is at 89°C. The dielectric losses at room temperature are $\tan\delta_\parallel = 0{,}01$ and $\tan\delta_\perp = 0{,}033$. The dielectric anisotropy at this temperature is 0,33, which is a comparable low value. But both components of losses are very high, especially the perpendicular component. At temperatures close to the clearing point these values are extremely high. Therefore this mixture is by no means suitable for microwave applications.

In Fig. 4.3.7 the permittivity and the losses, both perpendicular and parallel, versus temperature for the BMW13 mixture were presented.

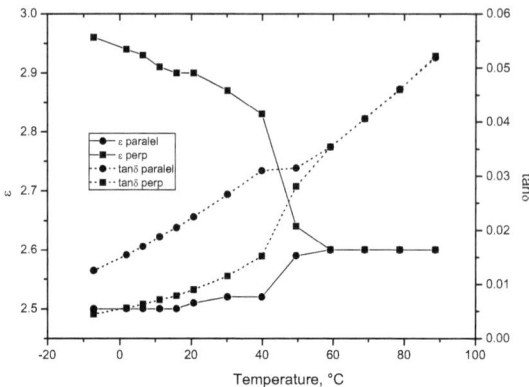

Fig. 4.3.7 Losses and permittivity versus temperature for BMW12

The losses at room temperature are $\tan\delta_\parallel = 0{,}009$, $\tan\delta_\perp = 0{,}022$, the dielectric anisotropy is 0,39. BMW13 shows noticeable better performance in comparison to BMW11. The reason should be the fact that BMW13 has quaterphenyl and terphenyl as additives (Add11,12), in comparison to BMW11 which has components with bridge fragments and hexane ring with remarkable lower values of optical birefringence. However it should be noted that still the value of the perpendicular component of losses is rather high and, because of the small breadth of the nematic phase, this mixture is a rather poor choice for microwave applications.

All other mixtures on the base of 5CB matrix show noticeable bad performance in the microwave region: low anisotropy and high values of losses. The reason for such tendency is, as we have found, the 5CB matrix by its own. The performance of the 5CB matrix has already left very few possibilities for improving the parameters of the mixture.
On the next Fig 4.3.8 one can see the properties of single 5CB compound at the frequency of 38GHz. One should note that the performance at 38 GHz should be better as at 30 GHz [Weil03].

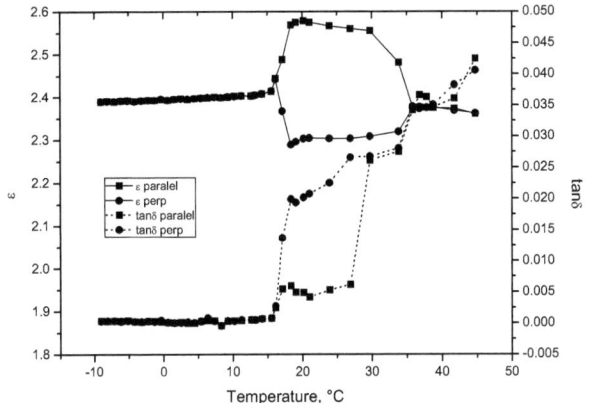

Fig. 4.3.8 Losses and permittivity versus temperature for 5CB

5CB possesses very small anisotropy and the values of losses are already rather high. The small temperature range of the nematic phase of 5CB also makes this material not suitable for microwave purposes.

4.3.3 Mixtures based on the Tolane matrix

On Fig. 4.3.9 the dependence of components of dielectric permittivity and losses with temperature for BMW8 is presented.

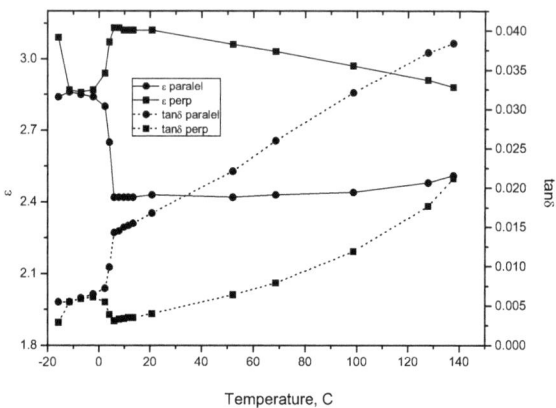

Fig. 4.3.9 Losses and permittivity versus temperature for BMW8

This mixture shows the same dependences as other investigated mixtures do, namely from the temperature. The losses at room temperature are $\tan\delta_\| = 0{,}004$, $\tan\delta_\perp = 0{,}017$, the dielectric anisotropy is 0,69. BMW8 shows a very good performance in comparison to mixtures based on the NCS or 5CB matrix. Both components of losses are noticeable smaller. At the same time because of the increased optical birefringence originated through the tolane matrix, the dielectric anisotropy at this frequency also has higher values.

On fig 4.3.10 the dependence of components of permittivity and losses for the mixture BMW10 is presented.

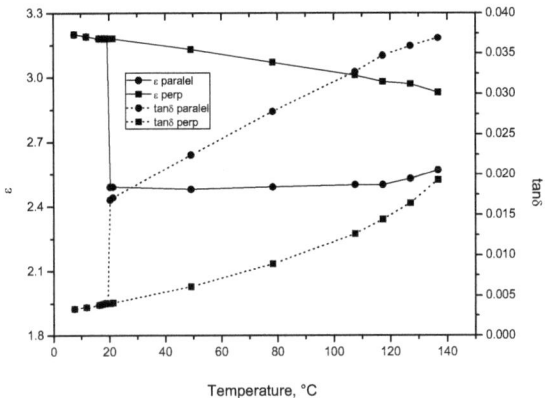

Fig. 4.3.10 Losses and permittivity versus temperature for BMW10

This mixture shows exactly the same dependences as BMW8. The losses at room temperature are $\tan\delta_\parallel = 0,004$, $\tan\delta_\perp = 0,017$, the dielectric anisotropy is 0,69. These values are equal to the values obtained for BMW8. However both mixtures have different dielectric anisotropy, even so the frequencies of the rotation around the short axis are different too. However it seems that these values don't affect the microwave performance directly. At the same time both mixtures have exactly the same value of optical birefringence, namely 0,33. Therefore the dielectric anisotropy at 30 GHz is the same for both mixtures.

Because of the same losses and the same dielectric anisotropy at GHz frequencies, the FOM parameter for both mixtures has similar behavior (fig. 4.3.11).

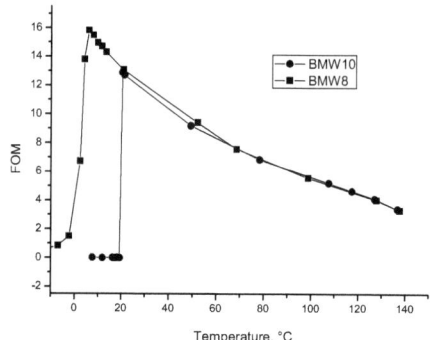

Fig. 4.3.11 Figure of Merit versus temperature for BMW8 and BMW10

The only difference lies in the lower transition temperature to the glassy or partial crystallized state of BMW8 which is below 0°C compared to BMW10 at 20°C.

As a result, the mixtures based on the tolane matrix are more preferable for any microwave application because of the higher birefringence of the tolane molecules.

4.4 Nematic mixtures at 38 GHz

Mixtures based on the tolane matrix were used and characterized in the microwave region at 38 GHz by the cavity perturbation method [Penirschke06] in a wide temperature range. The reason for selecting the fixed 38 GHz frequency compared to 30 GHz in 4.3 is based on the fact that with the increased frequency the performance of LC mixtures becomes better [Weil03].

The first investigated mixture is LHB13. As was shown previously (Chapter 3.5) this mixture is based on two matrixes and one additional quaterphenyl Comp1. The resulting birefringence is lower than those of the pure tolane matrix. Fig. 4.4.1 presents the components of permittivity and losses for LHB13.

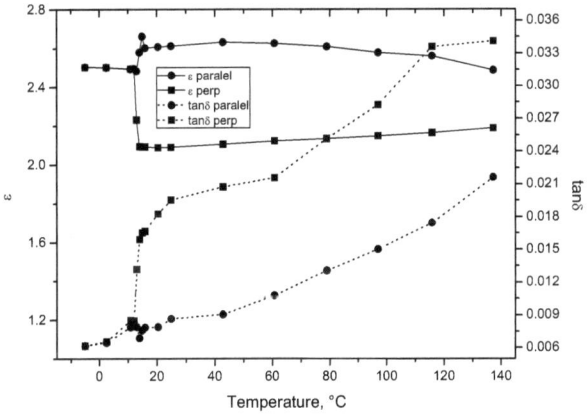

Fig. 4.4.1. Dielectric permittivity and losses for LHB13

The tanδ value for both components shows remarkable temperature dependence. The same dependences were achieved for mixtures of the BMW series at the frequency of 30GHz. The values of tanδ for parallel and perpendicular components at 20°C are 0,0079 and 0,0182, correspondingly. The dielectric anisotropy at this frequency is equal to 0,52. These values are in the same range as for the BMW mixtures, so we can see addition of one component Comp1 with high birefringence don't change the performance of this mixture. This can be explained by the fact that CompF1-3 have low value of optical birefringence.

The next investigated mixture LHB16 is based only on the tolane matrix and quaterphenyl additives (Comp2,3). The resulting birefringence is higher than for tolane matrix only. On fig. 4.4.2 the results of measurements for LHB16 mixture are presented.

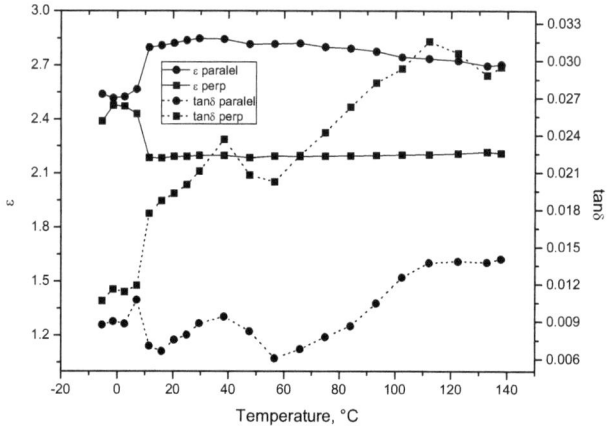

Fig. 4.4.2 Dielectric permittivity and losses for LHB16

This mixture shows rather different behaviour in comparison with the previous one (LHB13). We can see that below the point where both components of permittivity become equal, these components separate again. The reason can be the not full phase transition into the crystalline state or/and the forming of some kind of glassy phase. There is also a strong increase of losses in the range of 25-45°C The values of tanδ for parallel and perpendicular components at 20 C are 0,0075 and 0,0193, correspondingly. The dielectric anisotropy at this frequency is equal to 0,63. The losses are very close to values of LHB13; however the dielectric anisotropy is increased. The reason for that is the increased value of optical anisotropy of LHB16, because only the tolane matrix and quaterphenyls are present in this mixture, in contrary to LHB13 where matrix1 with Fluor atoms as lateral substituents was present.

LHB17 is based on the same matrix and Comp2,3 as LHB16 with only one difference namely the presence of additional quaterphenyl Comp4. On fig.4.4.3 one can see the results of the measurements for LHB17.

94

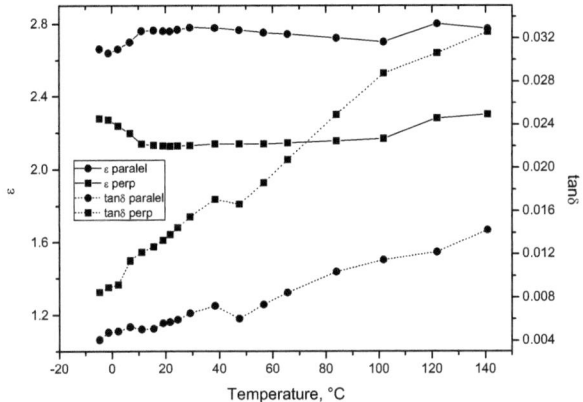

Fig. 4.4.3 Dielectric permittivity and losses for LHB17

This mixture shows better performance in comparison with LHB16, the values of losses are lower in the low temperature region. The values of tanδ for parallel and perpendicular components at 20°C are 0,0057 and 0,0138, correspondingly. The dielectric anisotropy at this frequency is equal to 0,63 like LHB16. This can be easily explained by the fact that both mixtures have the same values of optical birefringence. Because the only difference between LHB16 and LHB17 is the quaterphenyl Comp4, which has only 3% of weight percentage in LHB17, such great decrease of losses is rather unexpected. However sometimes this is true for the electro-optical properties of liquid crystals. This result shows that we can change the performance of mixtures in the microwave region already by addition of a small amount of specific compounds.

The mixture LHB18 is based on many quaterphenyl additives Comp1-5 and the tolane matrix. This mixture can be called a mixture with very high values of optical birefringence. On the fig. 4.4.4 the performance of this mixture at 38GHz is demonstrated.

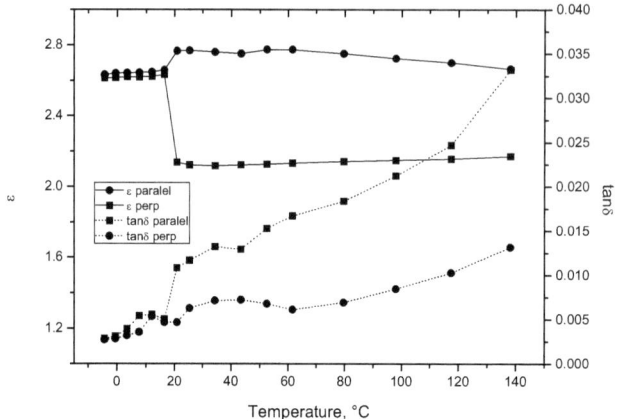

Fig. 4.4.4 Dielectric permittivity and losses for LHB18

LHB18 shows the best performance of all investigated mixtures. The losses are very small in the low temperature region. The values of tanδ for parallel and perpendicular components at 20°C are 0,0047 and 0,0108, correspondingly. The dielectric anisotropy at this frequency is equal 0,63 as for LHB17 and LHB16 because of very close values of the optical birefringence. The losses, especially the perpendicular components, are greatly decreased in comparison with other mixtures. This result shows that quarterphenyls with Fluor or NCS groups at the terminal position are good choices for the addition to the mixtures in order to decrease the losses and at the same time to reach to high value of dielectric anisotropy at microwave frequencies.

The Fig. 4.4.5 shows the Figure of Merit versus temperature for all LHB mixtures.

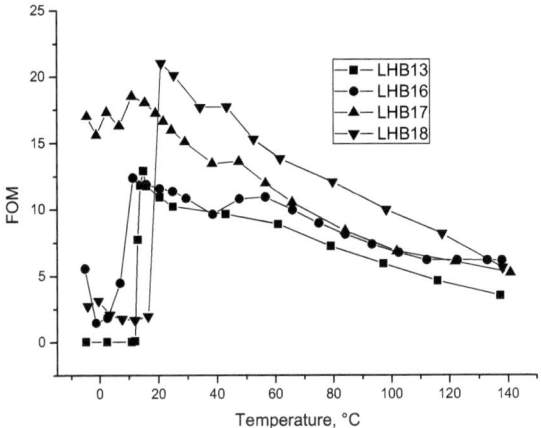

Fig. 4.4.5 FOM for LHB series of mixtures

As a result, LHB18 shows the best FOM value, at room temperature it's equal to 21, and this is a really very high value for liquid crystals.

The measurements in this chapter prove that the tolane matrix is very suitable for the preparation of mixtures for microwave purposes because of the high values of the optical anisotropy. Mixtures with a very broad range of nematic phase can be created. The only drawback of such compounds, the limited UV stability, is not very important by microwave experiments because usually such components like phase shifters are sealed and no day light can affect the materials. Furthermore is important, one can add quarterphenyls to the tolane mixture in order to improve as well the performance as the tunability, this is because of the high values of Δn.

4.5 Microwave performance of the investigated mixtures

So far the best performance in the microwave region can be observed with mixtures using a tolane base matrix. This one could expect because these compounds have also very high birefringence at optical frequencies. Furthermore favourable about these compounds is the wide temperature range of the nematic phase.

Out of the investigated base matrices, the next to the tolane matrix good performance is with the NCS matrix. In general, the used NCS compounds are not optimal because of the cyclohexane ring structure, which practically doesn't make any contribution to Δn of the molecules. In comparison, biphenyl NCS compounds should increase the performance of this matrix.

Mixtures on the base of 5CB LC compounds showed poor performance in our experiments. On table 4.5.1 the parameters of all investigated mixtures at room temperature and 30GHz are presented.

Table 4.5.1 Parameters of BMW series of mixtures.

Base Matrix	Mixture	$\varepsilon_\parallel$	$\tan\delta_\parallel$	$\varepsilon_\perp$	$\tan\delta_\perp$	τ	FOM	Δn	$\Delta\varepsilon$ (1kHz)
Tolane	BMW 10	3,05	0,004	2,39	0,017	0,22	12,51	0,33	6,25
Tolane	BMW 8	3,03	0,004	2,37	0,018	0,22	12,10	0,33	3,85
Tolane	BMW 7	3,10	0,004	2,43	0,019	0,22	11,56	0,33	8,42
Tolane	BMW 15	2,99	0,006	2,41	0,017	0,19	11,55	0,32	4,56
NCS	BMW 1	2,87	0,010	2,36	0,017	0,18	10,45	0,17	7,65
Tolane	BMW 5	3,07	0,005	2,40	0,022	0,22	10,15	0,30	0,22
NCS	BMW 4	2,95	0,088	2,36	0,021	0,20	9,35	0,21	7,51
NCS	BMW 2	2,90	0,010	2,41	0,019	0,17	8,89	0,17	9,69
NCS	BMW 3	2,88	0,011	2,40	0,022	0,17	7,75	0,18	10,81
5CB	BMW 13	2,86	0,010	2,49	0,024	0,13	5,39	0,27	11,61
5CB	BMW 14	2,84	0,010	2,48	0,026	0,13	4,89	0,26	12,85
5CB	BMW 12	2,83	0,015	2,53	0,032	0,11	3,42	0,22	10,49
5CB	BMW 16	2,86	0,010	2,54	0,034	0,11	3,31	0,23	8,5
5CB	BMW 11	2,81	0,010	2,50	0,034	0,11	3,23	0,25	11,33

As a result, different additives have different influence on the parameters of mixtures in the microwave region. However it is not easy to evaluate the influence of individual molecules in each mixture. On the other hand, the preparation of various mixtures was a need in order to receive some view on the best way to design mixtures.

The reason for the dielectric losses can be to some extent the tail of the described molecular relaxation around the short axis at about 1 MHz or higher. According to the Debye (or Cole-Cole) equation, the absorption tail can still be prominent at the frequencies up to 2-3 orders higher than the critical frequency of the certain process. For example, for the compound CCN5 because of the strong lateral dipolar CN group the diffuse reorientation around the long axis takes place at the frequency 110 MHz (was measured with the help of HP high frequency dielectric bridge). The Fig. 4.5.1 shows the Debye fit for this relaxation process.

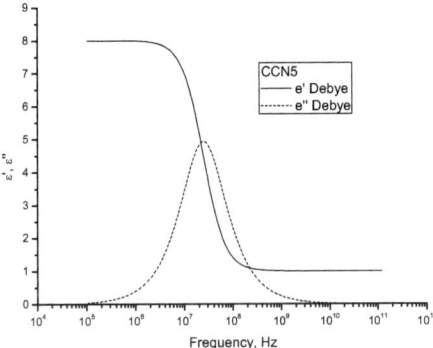

Fig. 4.5.1 Debye fit of the high frequency dispersion mode of CCN5

At frequencies up to 10 GHz the absorption tail is still present. One should note that in the case of non Debye like processes the absorption curve will broadened and therefore ε'' will be prominent even at higher frequencies. This was also pointed by [Utsumi04], where measurements of nematics at 1-10 GHz range were carried out. For mixtures with negative dielectric anisotropy and therefore strong dipole moment perpendicular to the long axis, the intensity of this dispersion mode at ca. 100 MHz or higher should be rather high in comparison with the molecular reorientation around the short axes by molecules with a strong longitudinal dipole moment. However the measurements of BMW21 [Chapter 3.5.6] shows that dielectric anisotropy at 30GHz and losses of this mixture have the values in the same range as other investigated mixtures. The anisotropy of BMW21 at 39°C is 0,51, tanδ perpendicular and parallel are rather high and equal to 0,356 and 0,11 correspondingly. That means the possible influence at ca. 30 GHz to this high frequency dispersion mode can't be very strong.

In [Utsumi04] several commercial nematic mixtures were investigated and it was stated that the mixture with higher values of $\Delta\varepsilon$ at low frequencies (close to static dielectric anisotropy) has the lowest losses in the microwave region. However results presented in this work shows that there is no indication for the direct influence of the dielectric anisotropy resulting in diffusive molecular reorientations at lower frequencies on the performance in the microwave region.

As a sideline of our work we calculated the vibration spectrum for some single molecules investigated with help of the Hyperchem software (v.7.0). It shows that there are possibilities for some complex vibrations around the frequencies of 150-200 GHz. However these processes can not influence 10^{10}Hz range. If such vibrations could greatly influence the losses in our frequency region the losses must raise with the increasing frequency. However experiments [Mueller05] shows that there is a decrease of losses up to 110 GHz for selected nematic mixtures.

There is a clear dependence between the quality factor and the birefringence of the mixtures (LHB) with a tolane matrix and quaterphenyl nematic components with a fluor atom or a NCS group at the terminal position. Those show the best quality factor in our investigations (Tab. 4.5.2).

Table 4.5.2 Parameters of LHB series of mixtures

Base Matrix	Mixture	$\varepsilon\|\|$	$\tan\delta\|\|$	$\varepsilon \perp$	$\tan\delta \perp$	τ	FOM	Δn	$\Delta\varepsilon$ (1kHz)
Tolane	LHB13	2,61	0,0083	2,18	0,0189	0,20	10,57	0,33	2,63
Tolane	LHB16	2,83	0,0077	2,19	0,0197	0,23	11,46	0,37	4,43
Tolane	LHB17	2,76	0,0058	2,13	0,0141	0,24	16,31	0,37	4,3
Tolane	LHB18	2,77	0,0055	2,13	0,0112	0,23	20,56	0,40	6,55

As result of our work some molecule structures have been indicated which perform a negative influence on the quality factor. For example some compounds with carboxylic (COO) or ethane (CH_2CH_2) bridging groups show high losses in the microwave region (30GHz). This can't be a high frequency residue of molecular reorientations because introducing a COO group or an ethane group in the molecules leads to higher viscosities and as usual a higher moment of inertia what means the relaxation frequency is shifted down towards lower frequencies. Alternatively one must assume the existence of low energy molecular vibrations, so called librations. Such librations of some dipolar groups could be expected at higher frequencies.

In another work the absorption process for selected LC compounds at frequencies around 100 cm^{-1} was observed [Janik83]. This process is mainly connected with librations of molecules around the long axis.

All in our work prepared mixtures were investigated over the broad nematic range. The experiments clearly show that the quality factor decreases with the increase of temperature. The reason for this is assigned to the decrease of the order parameter of the LC.

5 Summary

In this work dielectric and microwave investigations on different nematic liquid crystalline systems are presented. New knowledge about the influence of the chemical structure of molecules on the microwave performance was obtained. This information was used for the preparation of new, optimized nematic mixtures with high values of tunability and low dielectric losses in the microwave region. These new mixtures can be used for producing new devices like phase shifters or varactors. The dielectric measurements of investigated mixtures show that there is no direct influence of dielectric properties in the low frequency region on the GHz region.

The influence of different structural fragments of molecules on their properties in the dielectric and microwave region was studied with the help of prepared matrixes on the base of p-alkyl-cyclohexyl-phenyl-thiorhodanides, tolanes and 4-Cyano-4'-pentyl-biphenyl. All matrixes were chosen to have the nematic phase at room temperature. Single compounds that were used in this work allowed us to check the influence of different bridge groups, alkyl chains, polar groups, aromatic rings, on the properties of molecules.

The influence of the structure of chiral molecules with a SmC phase on the value of tilt angle was studied with the help of electro-optical and X-Ray diffraction methods. On the base of investigations of different quaterphenyl chiral compounds we were able to see how we can design molecules with high values of tilt angle. These data were used for preparation of new SmC* mixtures.

The new preliminary measurements of FLC (Ferroelectric Liquid Crystal) mixtures in the microstrip geometry were performed. So far, in literature there are only few reports on investigations of FLC's in coplanar and capacitance geometries. Our measurements showed that insertion losses for FLC mixtures are lower than for investigated nematic mixtures. That means there is possibility for using the FLC mixtures for devices where tunability is not that important, but very low values of losses are needed.

Dielectric measurements in the MHz region provide information about the low frequency molecular mode, assigned as reorientation around the short molecule axis, of prepared mixtures. It was shown that for almost all of the mixtures this mode has the values of critical frequency in the range of 100 kHz-5 MHz. Because the investigated single compounds have different shape and size, the separation of the absorption profile of the mixtures in individual contributions of the containing single molecules was possible. The obtained data shows that there is no any direct influence of the molecular mode, assigned as reorientation around the short molecular axis, on the properties in the GHz region. On the other hand reorientations around the long molecular axis taking place at up to 1 GHz might have some influence on the absorption processes in the microwave region. Within this work we could show only few examples because of the need of very strong lateral dipole moment on the investigated matrixes.

Nematic mixtures with a wide range of the nematic phase (>100°C), good anisotropy (high birefringence, $\Delta n=0.3-0.4$) and low losses in the microwave region were developed. So named LHB series of mixtures show good performance in both terms as are losses as well as quality factor and can be considered for use in real devices.

6 Zusammenfassung

In dieser Arbeit wurden dielektrische Untersuchungen und Untersuchungen im Mikrowellengebiet an verschiedenen nematischen flüssigkristallinen Systemen ausgeführt. Dabei konnten weitergehende Erkenntnisse über den Einfluss der chemischen Strukturen von Molekülen auf deren Verhalten im Mikrowellengebiet erhalten werden. Diese Informationen konnten für die Präparation von neuen, optimierten nematischen Mischungen mit hoher Steuerbarkeit und kleinen dielektrischen Verlusten im Mikrowellenbereich umgesetzt werden. Die neuen Mischungen können für die Herstellung von neuen Bauelementen wie Phasenschiebern oder Varactoren verwendet werden. Die dielektrischen Messungen an den untersuchten Mischungen haben gezeigt, dass aus den dielektrischen Eigenschaften im niedrigen Frequenzbereich kein direkter Einfluss auf die Eigenschaften im GHz Bereich ableitbar ist.

Der Einfluss von verschiedenen Strukturfragmenten am Molekül auf ihre dielektrischen Eigenschaften und ihr Verhalten im Mikrowellenbereich wurde mit Hilfe von präparierten Basismatrizen auf der Grundlage von p-Alkyl-cyclohexyl-phenyl-thiorhodaniden, Tolanen und 4-Cyano-4'-pentyl-biphenyl untersucht. Alle Matrizen wurden so eingestellt, dass die nematische Phase bei Zimmertemperatur existent ist. Die einzelnen Verbindungen, die in dieser Arbeit benutzt wurden, erlauben uns, den Einfluss von verschiedenen Brückengruppen, Alkylgruppen, polaren Gruppen und aromatischen Ringen auf das molekulare Eigenschaftsprofil hin zu untersuchen.

Der Einfluss der Struktur der chiralen Moleküle in einer SmC* Phase auf den Tiltwinkel wurde mit Hilfe von elektrooptischen und röntgenographischen Methoden untersucht. An verschiedenen chiralen Quaterphenylen ließ sich zeigen, wie Moleküle mit hohem Tiltwinkel konstruiert sein müssen. Diese Informationen wurden für die Präparation von neuen SmC* Mischungen eingesetzt.

Erste Untersuchungen an FLC (Ferroelectric Liquid Crystal)-Mischungen mittels ‚Microstrip Geometrie' konnten durchgeführt werden. Bisher wurde in der Literatur nur über wenige Arbeiten an FLC-Mischungen berichtet, meist in Koplanar- und Kapazitätsgeometrie. Unsere Untersuchungen zeigen, dass die Verluste an FLC-Mischungen kleiner als die bei den untersuchten nematischen Mischungen waren. Das bedeutet, dass FLC Mischungen für solche

Bauelemente, für die die Steuerbarkeit nicht im Vordergrund steht, dagegen kleine Verluste erzielt werden müssen, im Vorteil liegen.

Dielektrische Messungen im MHz Bereich liefern Informationen über die niederfrequente molekulare Mode der untersuchten Mischungen, die als Reorientierung um die kurze Molekülachse anzunehmen ist. Es wurde gezeigt, dass für fast alle Mischungen diese Mode eine kritische Frequenz im Bereich von 100 kHz-5 MHz hat. Aufgrund der verschiedenen Formen und Größen der untersuchten Moleküle konnte eine Auftrennung des Absorptionsprofils von Mischungen in die Einzelbeiträge der Komponenten erreicht werden. Die erhaltenen Daten zeigen, dass kein direkter Einfluss des Reorientierungsprozesses um die kurze Molekülachse auf das Absorptionsverhalten im Hochfrequenzbereich nachweisbar ist. Man darf aber davon ausgehen, dass der höherfrequente Reorientierungsprozeß um die lange Molekülachse, der bei bis zu 1 GHz liegt, im Mikrowellenbereich noch beitragen wird. Matrizes dieser Art müssen starke laterale Dipolmomente besitzen, die aber für diese Arbeit nur begrenzt zur Verfügung standen.

Mischungen mit einem breiten Bereich der nematische Phase (>100°C), guter Anisotropie (hoher Doppelbrechung, $\Delta n=0.3-0.4$) und niedrigen Verlusten im Mikrowellenbereich wurden entwickelt. Sogen. LHB-Serien von Mischungen zeigen sowohl gute Effizienz hinsichtlich der Verluste als auch einen hervorragenden ‚Qualityfactor'; diese Mischungen könnten für praktische Anwendungen eingesetzt werden.

7 Literature

[Belyaev99] B.A. Belyaev, N.A. Drokin, V.F. Shabanov, V.N. Shepov. *Physics of the Solid State.* **42**, 577 (1999)

[Bezborodov06] V. Bezborodov, V. Lapanik. *The 21st International Liquid Crystal Conference, Keystone, Colorado*, USA. **P84** (2006)

[Blinov94] L. M. Blinov, V. G. Chigrinov. *Electrooptic effects in liquid crystal materials.* Springer, New York, 481p (1994)

[Bose87] T.K.Bose, B. Campbell, S. Yagihara. *Phys. Review A.* **32**, 5767 (1987)

[Clark80] N. A. Clark, S. T. Lagerwall. *Appl. Phys. Lett.*, **36**, 899 (1980)

[Cole41] K.S. Cole, R.H. Cole. *J. Chem. Phys.*, **9**, 341 (1941)

[Daniel67] V.V. Daniel. *Dielectric Relaxation.* Academic Press, London, 281p (1967)

[Debye45] P. Debye. *Polar Molecules*, Dover Pubns., 172p (1945)

[Demus74] D. Demus, Ch. Fietkau, R. Schubert, H. Kehlen. *Mol. Cryst. Liq. Cryst.*, **25**, 215 (1974)

[Demus84] D. Demus, H. Zaschke. *Flüssige Kristalle in Tabellen II*, DVG Verlag, 468p, (1984)

[Dierking03] I. Dierking. *Textures of Liquid Crystals.* Wiley-VCH, 218p (2003)

[Dolfi93] D. Dolfi, M. Labeyrie, P. Joffre, J.P. Huignard. *Electronics Letters*, **29**, 926 (1993)

[Fujikake03] H. Fujikake, T.Kuki, H.Kamoda, F.Sato, T. Nomoto. *Appl. Phys. Lett*, **83**, 1815 (2003) [de Gennes93] P.G. de Gennes, J. Prost. *The Physics of Liquid Crystals*. Oxford Press, 616p (1993)

[Goelden07] F. Goelden, A. Lapanik, S. Mueller, A. Gaebler, W. Haase, R. Jakoby. *Microwave Conference 2007. European.* 106 (2007)

[Goelden07a] F. Gölden, A. Lapanik, A. Gaebler, Stefan Müller, Wolfgang Haase, Rolf Jakoby. *In IEEE LEOS Summer Topical Meeting.* (2007)

[Goodby98] Editors: J. Goodby, D. Demus, G.W. Gray, H.-W. Spiess, V. Vill. *Handbook of Liquid Crystals. Vol.2A.* Wiley-VCH (1998)

[Govers84] E. Govers, G. Vertagen. *Phys. Prew. A.* **30**, 1998 (1984)

[Grebenkin85] M. F. Grebenkin, V.F. Petrov, V.V.Belyaev. *Mol. Cryst. Liq. Cryst.* **129**, 245 (1985)

[Grebenkin89] M.F. Grebenkin, A.V. Ivascscenko. *Zhidkokristalicheskie Materialy*, 228p (1989)

[Gray89] G.W. Gray, M. Hird, D. Lacey, K.J. Toyne. *J. Chem. Soc.* Perkin Trans. 2, 2041 (1989)

[Gu07] M. Gu, Y. Yin, S. V. Shiyanovskii, O. D. Lavrentovich. *Phys. Rev. E.* **76**, 061702 (2007)

[Guerin97] F. Guerin, J.-M. Chappe, P. Joffre, D. Dolfi. *Jpn. J. App. Phys.* **36**. 4409 (1997)

[Haase03] Editors: W. Haase, S. Wrobel. *Relaxation Phenomena.* Springer, Berlin 748p (2003)

[Hanif06] B.M. Hanif, R.C. Hidalgo, D. E. Sullivan. *Phys. Rev. E.* **73**, 032701 (2006)

[Inui96] S. Inui, N. Iimura, T. Suzuki, H, Iwane, K. Miyachi, Y. Takanishi, A. Fukuda. *J. Mater. Chem.*, **6**, 671 (1996)

[Ivashchenko76] A.A. Ivashchenko, V.V. Titov, E.I. Kovshev. *Mol. Cryst. Liq. Cryst.* **33**, 195 (1976)

[Ivashchenko88] A.A. Ivashchenko, S.I. Torgova, L.A. Karamysheva, A.G. Abolin. *12th Intern. Liquid Crystal Conf. Progr. and Abstr., Freiburg. 1988*, P.97

[Jadzyn99] J. Jadzyn, S. Czerkas, G. Czechowski, A. Burczyk, R. Dabrowski. *Liq. Cryst.*, **3**, 437 (1999)

[Janik83] J.A. Janik, M. Godlewska, T. Grochulski, A. Kocot, E. Sciesinska, W. Witko. *Mol. Cryst. Liq. Cryst.* **98**, 67 (1983)

[de Jeu83] W.H. Jeu. *Phil. Trans. R. Soc. Lond.*, **A309**, 217 (1983)

[de Jeu88] W. H. de Jeu, G. Vertogen. *Thermotropic Liquid Crystals, Fundamentals*. Springer, Berlin, New York, 384p (1988)

[Kronig26] R. Kronig. *J.Opt.Soc.Am.*, **12**, 547 (1926)

[Kuki02] T.Kuki, H.Fujikake, T. Nomoto. *IEEE Trans. Microwave Theory Tech.* **50**, 2604 (2002)

[Maier57] W. Maier, A. Saupe. *Z. Naturforschung*, **12A**, 668 (1957)

[Maier61] W. Maier, G. Maier. *Z. Naturforschung*, **16A**, 262 (1961)

[Martin03] N. Martin, P. Laurent, G. Prigent, P. Gelin, F. Huret. "*European Microwave Conference (EuMC)*, Munich, (2003)

[Meyer75] R. B. Meyer, L. Liebert, L. Strzelecki, P. Keller. *Journ. Phys. Lett.*, **36**, (1975)

[Moritake05] H. Moritake, K. Toda, T. Kamei, Y. Utsumi, W. Haase. *Mol. Cryst. Liq. Cryst.* **434**, 527 (2005)

[Mueller04] S. Müller, P. Scheele, C. Weil, M. Wittek, C. Hock, R. Jakoby. *IEEE MTT-S Int. Microwave Symp.*, (2004)

[Mueller05] S. Mueller, A. Penirschke, C. Damm, P. Scheele, M. Wittek, C. Weil, R. Jakoby. *In IEEE Transactions on Microwave Theory and Techniques*, **53** 1937 (2005)

[Onsager49] L. Onsager. *Ann. N.Y. Acad. Sci.* **51**, 627 (1949)

[Osman81] M.A. Osman. *Mol. Cryst. Liq. Cryst.* **72**, 291 (1981)

[Osman85] M.A. Osman. *Mol. Cryst. Liq. Cryst.* **128**, 45 (1985)

[Penirschke06] A. Penirschke, S. Mueller, F. Goelden, A. Lapanik, V. Lapanik, V. Bezborodov, W. Haase, R. Jakoby. *GeMiC 2006, German Microwave Conference, Karlsruhe, Germany*, (2006)

[Pohl77] L. Pohl, R. Eidenschink, G. Krause, D. Erdman. *Phys. Lett.*, **60A**, 421 (1977)

[Pozar05] D.M. Pozar. *Microwave Engineering, Third Edition.* John Wiley & Sons, 720p (2005)

[Reinitzer88] F. Reinitzer. *Monatshefte für Chemie (Wien).* **9**, 421 (1888)

[Titov75] V.V. Titov, E.I. Kovshev, A.I. Pavluchenko. *J. Phys.* **46**, 387 (1975)

[Utsumi05] Y. Utsumi, T. Kamei. *Mol. Cryst. Liq. Cryst.* **434**, 329 (2005)

[Utsumi04] Y. Utsumi, T. Kamei. *Mol. Cryst. Liq. Cryst.* **409**, 355 (2004)

[Utsumi05a] Y. Utsumi, T. Kamei, R. Naito, K. Saito. *Mol. Cryst. Liq. Cryst.* **434**, 337 (2005)

[V.Lapanik04] V.Lapanik, V.Bezborodov, W.Haase. *Book of Abstracts, 20th International Liquid Crystal Conference, Ljubljana, Slovenia*, **P-020** (2004)

[Watson02] S.J Watson, L.S. Matkin, L.J. Baylis, N. Bowring, H.F. Gleeson, M. Hird, J. Goodby. *Phys. Rev. E.*, **65**, 03175 (2002)

[Weil02] C. Weil, G. Lüssem, R. Jakoby. *IEEE MTT-S Int. Microwave Symp.*, 367, (2002)

[Weil03] C. Weil, S. Müller, P. Scheele, Y. Kryvoshapka, P. Best, G. Lüssem, R. Jakoby. *In Proc. 33th European Microwave Conf.*, **3**, 1431 (2003)

[Weil03a] C. Weil, St. Müller, P. Scheele, P. Best, G. Lüssem, R. Jakoby. *Electronics Letters*, **39**, 1732 (2003)

Die VDM Verlagsservicegesellschaft sucht für wissenschaftliche Verlage abgeschlossene und herausragende

Dissertationen, Habilitationen, Diplomarbeiten, Master Theses, Magisterarbeiten usw.

für die kostenlose Publikation als Fachbuch.

Sie verfügen über eine Arbeit, die hohen inhaltlichen und formalen Ansprüchen genügt, und haben Interesse an einer honorarvergüteten Publikation?

Dann senden Sie bitte erste Informationen über sich und Ihre Arbeit per Email an *info@vdm-vsg.de*.

Sie erhalten kurzfristig unser Feedback!

VDM Verlagsservicegesellschaft mbH
Dudweiler Landstr. 99　　　　　　　Telefon +49 681 3720 174
D - 66123 Saarbrücken　　　　　　　Fax　　 +49 681 3720 1749
www.vdm-vsg.de

Die VDM Verlagsservicegesellschaft mbH vertritt

Printed by Books on Demand GmbH, Norderstedt / Germany